Compact Textbooks in Mathematics

This textbook series presents concise introductions to current topics in mathematics and mainly addresses advanced undergraduates and master students. The concept is to offer small books covering subject matter equivalent to 2- or 3-hour lectures or seminars which are also suitable for self-study. The books provide students and teachers with new perspectives and novel approaches. They may feature examples and exercises to illustrate key concepts and applications of the theoretical contents. The series also includes textbooks specifically speaking to the needs of students from other disciplines such as physics, computer science, engineering, life sciences, finance.

- **compact:** small books presenting the relevant knowledge
- **learning made easy:** examples and exercises illustrate the application of the contents
- **useful for lecturers:** each title can serve as basis and guideline for a semester course/lecture/seminar of 2-3 hours per week.

Siddhartha Pratim Chakrabarty ·
Ankur Kanaujiya

Mathematical Portfolio Theory and Analysis

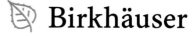 Birkhäuser

Siddhartha Pratim Chakrabarty🆔
Department of Mathematics
Indian Institute of Technology Guwahati
Guwahati, Assam, India

Ankur Kanaujiya🆔
Department of Mathematics
National Institute of Technology Rourkela
Rourkela, Odisha, India

This textbook has been reviewed and accepted by the Editorial Board of Mathematik Kompakt, the Germany language version of this series.

ISSN 2296-4568   ISSN 2296-455X   (electronic)
Compact Textbooks in Mathematics
ISBN 978-981-19-8543-0   ISBN 978-981-19-8544-7   (eBook)
https://doi.org/10.1007/978-981-19-8544-7

Mathematics Subject Classification: 49L12, 49L20, 60G15, 62P05, 91-01, 91-10, 91G10, 91G70

This book is published under the imprint Birkhäuser, www.birkhauser-science.com by the registered company Springer Nature Singapore Pte Ltd.
The registered company address is: 152 Beach Road, #21-01/04 Gateway East, Singapore 189721, Singapore

Preface

The transition from the traditional commercial banking activities of lending and borrowing, to the modern-day financial structure involving risky securities and derivatives, has resulted in the necessity of active management of assets by qualified professionals, both in terms of investment strategies and in terms of the consequent risk management of these investments.

This book is a result of the first author's (Siddhartha Pratim Chakrabarty) teaching of an undergraduate elective on portfolio theory and performance analysis, to the final-year students of the Bachelor of Technology (in Mathematics and Computing) at the Indian Institute of Technology (IIT) Guwahati. During the period of this course being offered for more than a decade, the necessity of a book encompassing a wide spectrum of mathematical portfolio theory and analysis at an introductory level was strongly felt. The goal was to develop a book which provides a holistic insight into the topic at an undergraduate level. While the final-year undergraduate students, who were taught this course at IIT Guwahati, had a strong background in financial engineering and stochastic calculus, this book requires no such prerequisites, for the intended audience. Accordingly, the book begins with chapters on financial markets, basic probability theory, and pricing models, as a prelude, before embarking on the discussion on portfolio theory. The emphasis at the commencement is on the modern portfolio theory (or the mean-variance portfolio theory) due to Harry Markowitz, followed by two chapters, one on utility theory and the other on non-mean-variance portfolio theory.

The next two chapters are on topics typically not covered in an undergraduate text on mathematical finance. The former is on optimal portfolios, both in discrete and continuous time setup, via the dynamic programming principle, and the Hamilton–Jacobi–Bellman equation, respectively, and the latter is on optimization of bond portfolios, introduced to make the reader aware of the importance of management of bond portfolios (contrary to the perception that bonds are "risk-free" and hence require little or no active management). The concluding chapter deals with a risk management technique, namely Value-at-Risk (VaR), which is playing a progressively important role, especially in compliance with the capital requirements under the Basel Accord. The second author (Ankur Kanaujiya), as a graduate student at IIT Guwahati, has worked in the area of computational aspects of financial derivatives, with a trading account as the underlying asset.

The first author acknowledges his students at IIT Guwahati (a group of highly gifted young men and women) who have provided their feedback, on the course, over the years. That has, in many ways, shaped the structure of the book. Finally, in summary, the purpose of writing this introductory book is to achieve a textbook, that the authors themselves would have like to have, as students of this ever-evolving subject.

Guwahati, India   Siddhartha Pratim Chakrabarty
Rourkela, India   Ankur Kanaujiya
October 2022

Contents

About the Authors

Siddhartha Pratim Chakrabarty is Professor at the Department of Mathematics, Indian Institute of Technology Guwahati, Assam, India. With a long and varied experience of teaching and undertaking research work in a wide spectrum of subareas of finance, including financial engineering, mathematical finance, computational finance, Monte–Carlo simulation, portfolio theory and financial risk management, he has offered two massive online open course (MOOC) courses through the National Program for Technology Enhanced Learning (NPTEL), on mathematical finance and mathematical portfolio theory. Professor Chakrabarty is very passionate about undergraduate research, which has led to several publications with his undergraduate students, many of whom have gone on to secure prestigious positions in academia, data science, entrepreneurship and investment banking. In 2020, he was a recipient of the Scholarship Scheme for Faculty Members from Academic Institutions 2020 by the Reserve Bank of India. A very active in professional services and outreach activities, Prof. Chakrabarty has supervised four Ph.D. students and published 46 research articles.

Ankur Kanaujiya is Assistant Professor at the Department of Mathematics, National Institute of Technology Rourkela, Odisha, India. After completing his Ph.D. in the area of computational finance, he joined Birla Institute of Technology Mesra, Ranchi, Jharkhand, India, under the Technical Education Quality Improvement Programme (TEQIP) III before he moved to his current position. Dr. Kanaujiya has worked extensively in the areas of computational and applied mathematics with an emphasis on finance.

List of Figures

Mechanisms of Financial Markets

1

The diversity and volume of trading in financial markets have seen a dynamic and continuous evolution over the last few decades. In this introductory chapter, we emphasize on three key components of market mechanism, namely the types of markets, the players operating in these markets, and finally, the basic types of financial instruments which are traded in such markets.

1.1 Types of Markets

At a global level, the financial markets can be subdivided into what are known as local (or onshore) and what is known as Euromarkets. The two key characteristics of onshore markets are the formal registration process and the requirements of maintaining reserves, overseen by regulatory agencies and central banks, which leads to implications, such as cost, liquidity, and taxation. In case of the latter, namely the Euromarkets, both these characteristics, i.e., the registration process and the reserves are beyond the jurisdiction of regulators and central banks.

The onshore markets can be classified as either being exchange traded or Over-the-Counter (OTC) markets. For example, a derivatives (to be defined later in the chapter) exchange is a market where standardized contracts (designed by the exchange) are traded. Exchanges traditionally used what is known as the "open-outcry-system", with the traders physically meeting on the floor of the exchange, but have gradually moved on to the "electronic system". Chicago Board of Trade (CBOT) and the Chicago Mercantile Exchange (CME) are two early derivatives exchanges, set up in 1848 and 1919, respectively, to primarily trade in futures contracts (to be defined later in the chapter). This was followed by the commencement of trading of call and put options at the Chicago Board of Options Exchange (CBOE) in 1973 and 1977, respectively. In addition, there are numerous stock exchanges, which are exchanges dealing with the trading of equities. On the other hand, the OTC markets

S. P. Chakrabarty and A. Kanaujiya, *Mathematical Portfolio Theory and Analysis*, Compact Textbooks in Mathematics, https://doi.org/10.1007/978-981-19-8544-7_1

provide for an important (and more flexible) alternative to exchanges. In practice, OTC markets are much larger than the exchange traded markets, in terms of the total volume of trading. The OTC trading is carried out over an electronic network or telephone and typically takes place between two financial institutions or between a financial institution and one of its clients. While OTC markets have more flexibility, in terms of negotiation of individualized contracts, such contracts, nevertheless, are overseen by regulatory authorities, with the contract documentation being prepared by professional associations. As already noted, the Euromarkets (with, of course, no connection to Europe, in particular) provides for trading an instrument without formal registration and less regulatory oversight. An example of this would be a security traded without registration and called a bearer's security.

1.2   Market Players

The functioning of markets involves several key players, such as traders, brokers, risk managers, and analysts. Traders are individuals trading in financial instruments and can be broadly classified into three categories, namely hedgers, speculators, and arbitrageurs. Brokers provide an intermediary platform for the buyers and sellers to carry out their trade, through a bid and ask process. The enhanced level of risk resulting from the emergence of sophisticated financial instruments has led to the involvement of risk managers, who assess the potential risk of a trade and whose approval is required prior to carrying out such trades. Finally, analysts are researchers, whose primary role is to carry out a detailed analysis of instruments and provide advisory services, while having no role in the trading process itself.

1.3   Financial Instruments

In this section, we focus on the concept of financial instruments, with an emphasis on the economic use of such instruments. Accordingly, we consider two very broad categories of such instruments, namely securities and financial derivative contracts. In the case of securities, we consider two basic securities, namely fixed income and equities, with a particular emphasis on bonds and stocks, respectively, essentially as a prelude to their extensive consideration for our subsequent discussion on portfolio theory. In addition, we also elaborate on three key financial derivative contracts, namely futures and forwards, options, and swaps, mainly from the perspective of their financial mechanism.

1.3.1 Bonds

A "bond" is a security, that is issued by a borrower (selling a bond) to the lender (purchasing a bond), in lieu of a specified amount of borrowing (the bond price). The bond confers the right to the lender (and a consequent obligation on the borrower) to receive a single (or a series of) pre-specified payments at future pre-specified date(s). The borrower or the "seller of the bond" is called the "debtor", whereas the lender or the "purchaser of the bond" is called the "creditor". Accordingly, the debtor and the creditor are called the "counterparty" of each other. In case a single payment is made by the creditor to the debtor, then the bond is called a "zero-coupon bond". However, if the bond involves a series of payments, then it is called a "coupon bond". The single payment made in case of a zero-coupon bond is called the "par value", "face value", "nominal value", or "principal" (henceforth, in order to avoid any confusion, we will make use of the term "par value", in this context). The series of payments made to the owner of a coupon bond are called "coupons", with the final payment comprising of the last coupon and the par value. The date on which the par value is paid, in case of bonds, is called the "maturity" of the bond. Recall that the amount paid by the purchaser of the bond to the seller of the bond is called the "bond price". In case of a bond, the difference between the bond price and the par value as a percentage of the bond price is called the "interest rate". In case of a coupon bond, the periodic coupons (typically paid semi-annually) are usually a pre-specified percentage of the par value and are called the "coupon rate". In summary, the par value, the coupon rate, and the maturity characterize the contract between the seller and the purchaser, of the bond, and are referred to as the "bond indenture".

In principle, a bond will represent risk-free security, in the sense that it offers guaranteed future payment(s), that are known in advance, to the purchaser of the bond, at the time of its purchase. However, if one looks beyond this aspect of guaranteed payments being known in advance, bonds might actually involve several types of risks. In the event that the issuer of the bond fails to meet the payment obligations to the purchaser of the bond, then the issuer is said to have defaulted, resulting in loss for the purchaser. This kind of risk faced by the bond owner is known as "credit" or "default" risk. Also, the case of the rate of increase in the value of goods, being more than the rate of return received from a bond, results in diminishing purchasing power for the bond owner, resulting from what is known as the "inflation" risk. Finally, given the fact that bonds can be liquid, it is possible that the bond owner may sell the bond to someone, between the purchase date and the maturity of the bond, but end up receiving an amount for the bond that is lower than what they had anticipated at the time of purchase (this typically happens if the interest rate goes up). This kind of risk arising from pre-mature (prior to maturity) liquidation of the bond is known as the "liquidity" risk.

1.3.2 Stocks

A (common) stock or equity is a security that bestows upon the shareholder, ownership shares of the company (or corporation) that has issued the stocks, and gives the right to the proportion of any profits earned by the company, that it decides to distribute among the shareholders (instead of reinvesting them in the company), the proportional part of the company in the event of its closure, and consequent liquidation. The shareholder of the (common) stock is entitled to one vote per (common) stock during the voting on agendas placed in the company's annual general meeting. The profits that the company distributes to its (common) stockholders are referred to as dividends, which (unlike coupons and par value, in case of bonds) are random and are not known in advance. In fact, the randomness or uncertainty of dividends is the primary distinguishing feature of (common) stocks, via-a-vis bonds. Like bonds, stocks are liquid, i.e., the stockholder can sell the stock at any time. This could result in a profit or loss (depending on whether the selling price is more than or less than the purchase price) and is called capital gains or loss, respectively. Thus the return of a (common) stock has two components, namely dividend yield and capital gains/loss.

In contrast, a (preferred) stock has features akin to that of a bond, in that, it promises the owner a fixed amount of return on an yearly basis, in principle acts like a perpetual bond, and comes with no voting privileges about the management of the company. The company, at its discretion, may defer the dividend payments to preferred stockholders, and such payments are usually cumulative. However, dividends must first be paid to the (preferred) shareholders, before any dividend payment is made to the (common) shareholders. In the event of a bankruptcy and consequent liquidation, the (preferred) stockholders get precedence over (common) shareholders, in the payout of the recovered amount.

We end the narrative on equities (and by this we will refer to common stock in this context) by discussing the concept of going long or short on a stock. Suppose that an investor believes that a stock is overvalued, but is not in ownership of the stock. Then the investor can take a position, driven by their belief, through a process called short-selling. The concept of short-selling a stock involves the borrowing of the stock from its owner, and then selling it in the market (short-selling), based on the projection that the stock is overvalued and would see a fall in its price, which then can be purchased by the short-seller and returned to the owner, an act that is called "covering the short position". In contrast, the act of purchasing and consequent ownership of stock is called the long position in the stock.

1.3.3 Derivatives

Financial derivatives (as the name suggests) are financial instruments, whose value depends on or is "derived" from the value of an underlying asset, such as prices of bonds and stocks, interest rates, market indices. These are sometimes referred to as "contingent claims", since their payoff is dependent or "contingent" on the values of the underlying assets. We will focus our discussion on three fundamental types of financial derivatives, namely forwards and futures and options and swaps.

1. *Forwards and futures:* A forward contract (or forwards) is an agreement between two parties, to buy and sell an underlying asset, at a pre-specified future date, for a pre-specified price. The pre-specified future date is called the "maturity", and the pre-specified price is called the "forward price". This is in contrast to the "spot-price", which is the price at which the purchase of the underlying asset can be made "on the spot". The party in the forward contract, which agrees to buy the underlying asset, is said to have the "long position", whereas its counterparty, which agrees to sell the asset, is said to have the "short position". Forwards are traded in OTC markets and are legally binding on both the parties.

   Like a forward contract, a futures contract (or futures) is an agreement between two parties, to buy and sell an underlying asset at a pre-specified future date (maturity) for a pre-specified price (futures price). However, unlike forwards, the futures are traded on exchanges, which specifies a standardized futures contract and provides for a mechanism called marking-to-market, which is designed to ensure that any default by one party of this legally binding futures contract does not cause loss to the counterparty.

2. *Options:* Options are agreements, wherein one of the parties has the right (called the holder of the option) to buy (sell), with the counterparty having the obligation (called the writer of the option) to sell (buy), an underlying asset, on or before a pre-specified future date (exercise time or expiration date or maturity) for a pre-specified price (exercise price or strike price). Unlike forwards and futures, options are only legally binding on one of the parties of the agreement. We now look at the classification of options, from two perspectives, namely the party with the legally binding position, and whether the option is exercised on or before the expiration date.

   Depending on the party on whom the option agreement is legally binding, we can classify options as call and put options. A "call option" gives the holder of the option, the right (but not the obligation) to buy the underlying asset for a pre-specified maturity and expiration date. Accordingly, the obligation (but not the right) to sell the underlying asset lies on the writer of the option, for which the writer of the call option receives an upfront premium, known as the "price" of the call option, from the holder of the call option. In an analogous manner, a "put option" gives the holder of the option the right (but not the obligation) to sell the underlying asset for a pre-specified maturity and expiration. Accordingly, the obligation (but not the right) to sell the underlying asset lies on the writer of the option, for which the writer of the put option receives an upfront premium, known as the "price" of the put option, from the holder of the put option.

   Another classification of options can be made on the basis of when the option can be exercised. An option which can be exercised only at the expiration date is called an European option. In contrast, an option, which allows for the flexibility of being exercised at any time, on or before the expiration date is called an American option. Clearly, the holder of an American option has greater flexibility than the holder of an European option, which renders the price of an American option to be more than that of an European option.

3. *Swaps:* A swap is an agreement between two parties to exchange cash flows in the future. Two commonly used swaps are interest rate swaps and currency swaps. An interest rate swap is an agreement in which the interest rates are swapped on a "notional principal", which is the same for both the parties. In such a swap, typically, a fixed interest rate is exchanged or swapped for a floating interest rate on a notional principal. Also, quite frequently, interest rates on bonds of different currencies differ from each other, and this could lead to not only the exchange of interest rates, but also the currencies. In case only interest rates are exchanged, it is called "interest rate swaps", but when the interest rate and currencies are both exchanged, it is known as the "currency swaps". Of course, the exchange could involve market variables, other than interest rates and currencies. The swap agreement will specify the dates on which the exchanges take place and the mechanism of calculating them.

Fundamentals of Probability Theory

2

The study of portfolio theory involves the prediction of future events, particularly the estimation of future values of assets, which inherently is random in nature. Accordingly, given the different possible outcomes, this area is heavily reliant on tools from probability theory. At the heart of this theory lies the concept of outcomes of an experiment and the likelihood or probability of each of those outcomes. The set of all possible outcomes is called the *sample space*, which may either be *discrete*, a term which includes finite or countably infinite sample spaces, or *continuous* sample spaces. Accordingly, we first discuss the probability theory for finite sample spaces and then for general (countably infinite and continuous) sample spaces.

2.1 Finite Probability Space

Let us consider an experiment, whose finite sample space is denoted by S. If A is a subset of the sample space S, then it is called an event. If A_1 and A_2 are two events in S, then their union and intersection are also events, denoted by $A_1 \cup A_2$ and $A_1 \cap A_2$, respectively. $A_1 \cup A_2$ is an event whose outcomes lie in either A_1 or A_2 or both, whereas $A_1 \cap A_2$ is an event whose outcomes lie in both A_1 and A_2. The events A_1 and A_2 are said to be mutually exclusive, if $A_1 \cap A_2 = \phi$, where ϕ is the null event. The event A^c is the complement of the event A which comprises of outcomes that are not in A. If S is a finite sample space, then for each $s_i \in S$, the event $A_i^* = \{s_i\}$ is called an elementary event.

Definition 2.1.1 (*Finite Probability Space*). Let S be a finite sample space and let $\mathbb{P}$ be a real-valued function defined on all the events in S, with $\mathbb{P}$ (called the probability measure) satisfying the following properties:

(i) For all events $A \subseteq S$, $0 \leq \mathbb{P}(A) \leq 1$.
(ii) For the sample space S, $\mathbb{P}(S) = 1$.

© The Author(s), under exclusive license to Springer Nature Singapore Pte Ltd. 2023
S. P. Chakrabarty and A. Kanaujiya, *Mathematical Portfolio Theory and Analysis*,
Compact Textbooks in Mathematics, https://doi.org/10.1007/978-981-19-8544-7_2

(iii) If the events $A_1, A_2, \ldots, A_n$ are mutually exclusive, i.e., $A_i \cap A_j = \phi$ for $i, j = 1, 2, \ldots, n, i \neq j$, then $\mathbb{P}\left(\bigcup_{i=1}^{n} A_i\right) = \sum_{i=1}^{n} \mathbb{P}(A_i)$.

Then the pair $(\mathcal{S}, \mathbb{P})$ is called a finite probability space.

The following are the consequence of the above definition:

(i) If ϕ is the null event, then $\mathbb{P}(\phi) = 0$.
(ii) If $A_1 \subseteq A_2$, then $\mathbb{P}(A_1) \leq \mathbb{P}(A_2)$.
(iii) For an event A, $\mathbb{P}(A^c) = 1 - \mathbb{P}(A)$.

Definition 2.1.2 (*Probability Mass Function*). The probability mass function is a real-valued function $f : \mathcal{S} \to \mathbb{R}$ defined as $f(s_i) = p_i$, where $p_i = \mathbb{P}(A_i^*)$. Note that the set $\{p_i | s_i \in \mathcal{S}\}$ is called the probability distribution.

It follows immediately that, $\mathbb{P}(A) = \sum_{s_i \in A} \mathbb{P}(A_i^*)$.

Definition 2.1.3 (*Independence of Events*). The events $A_1, A_2, \ldots, A_n$ are said to be independent, if for all subcollection $\{A_{i_1}, A_{i_2}, \ldots, A_{i_m}\}$, $\mathbb{P}\left(\bigcup_{k=1}^{m} A_{i_k}\right) = \prod_{k=1}^{m} \mathbb{P}\left(A_{i_k}\right)$.

Definition 2.1.4 (*Random Variable*). Let $(\mathcal{S}, \mathbb{P})$ be a finite probability space. Then a real-valued function $X : \mathcal{S} \to \mathbb{R}$ is called a random variable on the finite sample space $\mathcal{S}$.

Now, suppose that the random variable takes the discrete values $\{x_1, x_2, \ldots, x_m\}$, then for each $s_i \in \mathcal{S}$, the random variable X takes one of the values from $\{x_1, x_2, \ldots, x_m\}$, that is $X(s_i) = x_k$ for some $k = 1, 2, \ldots, m$. Then the collection of all such s_i, for which $X(s_i) = x_k$ is the event $\{X = x_k\} = \{s_i | X(s_i) = x_k\}$, which is denoted as $X^{-1}(x_k)$. On the other hand, if X takes a continuous set of values, then equivalently, we have $\{X \leq x_k\} = \{s_i | X(s_i) \leq x_k\}$, which is denoted as $X^{-1}((-\infty, x_k))$. This brings us again to the definition of the probability distribution, but this time in the context of the random variable.

Definition 2.1.5 (*Probability Distribution for a Random Variable*). The probability distribution or probability measure for the random variable X is defined as $\mathbb{P}_X(\{x_k\}) = \mathbb{P}(\{X = x_k\})$ or as $\mathbb{P}_X(\{x_k\}) = \mathbb{P}(\{X \leq x_k\})$, for the discrete or continuous set of values taken by the random variable X, respectively.

Definition 2.1.6 (*Independence of Random Variable*). The random variables X_1, X_2, ..., X_n are said to be independent if their corresponding events denoted by $X_1 = x_1$, $X_2 = x_2, \ldots, X_n = x_n$ are independent, that is, $\mathbb{P}(X_1 = x_1, X_2 = x_2, \ldots, X_n = x_n) = \prod_{i=1}^{n} \mathbb{P}(X_i = x_i)$.

Definition 2.1.7 (*Expectation*). Let X be a random variable defined on the finite probability space $(S, \mathbb{P})$. Then the expectation (or expected value) of a random variable $X : S \to \mathbb{R}$ is defined as, $\mu_X = E(X) = \sum_{i=1}^{n} X(s_i)\mathbb{P}(s_i)$. In particular, if X takes the values from $\{x_1, x_2, \ldots, x_m\}$, then $\mu_X = E(X) = \sum_{i=1}^{m} x_i \mathbb{P}(X = x_i)$.

Some of the properties of expectation are enumerated below:

1. If $X_1, X_2, \ldots, X_n$ are random variables, and $a_1, a_2, \ldots, a_n$ are constants, then,

$$E\left(\sum_{i=1}^{n} a_i X_i\right) = \sum_{i=1}^{n} a_i E(X_i).$$

2. If the function $f : \mathbb{R} \to \mathbb{R}$ is a real-valued function and X is a random variable, then the composite function $f(X) : S \to \mathbb{R}$ is also a random variable, whose expectation is given by,

$$E(f(X)) = \sum_{i=1}^{n} f(X(s_i))\mathbb{P}(s_i) = \sum_{i=1}^{m} f(x_i)\mathbb{P}(X = x_i).$$

3. If $X_1, X_2, \ldots, X_n$ are independent random variables, then,

$$E\left(\prod_{i=1}^{n} X_i\right) = \prod_{i=1}^{n} E(X_i).$$

Definition 2.1.8 (*Variance and Standard Deviation*). Let X be a random variable defined on the finite probability space $(S, \mathbb{P})$, with finite expected value μ_X. Then the variance of the random variable X is defined as, $\sigma_X^2 = \text{Var}(X) = E\left[(X - \mu_X)^2\right]$. Further the standard deviation of the random variable X is defined as, $\sigma_X = SD(X) = \sqrt{\text{Var}(X)}$. Note that the standard deviation is the positive square root of the variance.

Definition 2.1.9 (*Covariance*). Let X and Y be a random variable defined on the finite probability space $(S, \mathbb{P})$, with respective finite expected values μ_X and μ_Y. Then, the covariance of the random variable X and Y is defined as, $\sigma_{XY} = \text{Cov}(X, Y) = E\left[(X - \mu_X)(Y - \mu_Y)\right]$.

Some of the properties of variance and covariance are enumerated below:

1. If X is a random variable, then,

   (a) $\sigma_X^2 = \text{Var}(X) = E(X^2) - \mu_X^2 = E(X^2) - (E(X))^2$.
   (b) For $a \in \mathbb{R}$, $\text{Var}(aX) = a^2\text{Var}(X)$.
   (c) For $a \in \mathbb{R}$, $\text{Var}(X + a) = \text{Var}(X)$.

2. If $X_1, X_2, \ldots, X_n$ are independent random variables, and $a_1, a_2, \ldots, a_n$ are constants, then,

$$\text{Var}\left(\sum_{i=1}^{n} a_i X_i\right) = \sum_{i=1}^{n} \sum_{j=1}^{n} a_i a_j \text{Cov}(X_i, X_j).$$

3. If X and Y are random variables, then,

   (a) $\text{Cov}(X, Y) = E(XY) - E(X)E(Y)$.
   (b) $\text{Cov}(X, Y) = \text{Cov}(Y, X)$.
   (c) $\text{Cov}(X, X) = \sigma_X^2$.
   (d) If $X = a_1$ or $Y = a_2$ or both, for $a_1, a_2 \in \mathbb{R}$, then $\text{Cov}(X, Y) = 0$.
   (e) $|\text{Cov}(X, Y)| \leq \sigma_X \sigma_Y$, with the equality holding if and only if, either X or Y is constant, or if $Y = a_1 X + a_2$ for $a_1, a_2 \in \mathbb{R}$.

4. If $X_1, X_2, \ldots, X_n$ and Y are random variables then,

$$\text{Cov}\left(\sum_{i=1}^{n} a_i X_i, Y\right) = \sum_{i=1}^{n} a_i \text{Cov}(X_i, Y).$$

5. If $X_1, X_2, \ldots, X_n$ and $Y_1, Y_2, \ldots, Y_m$ are random variables then,

$$\text{Cov}\left(\sum_{i=1}^{n} a_i X_i, \sum_{j=1}^{m} b_j Y_j\right) = \sum_{i=1}^{n} \sum_{j=1}^{m} a_i b_j \text{Cov}(X_i, Y_j).$$

Definition 2.1.10 (*Correlation Coefficient*). Let X and Y be a random variable defined on the finite probability space $(S, \mathbb{P})$, with respective finite expected values μ_X and μ_Y and respective finite non-zero variances σ_X^2 and σ_Y^2. Then the correlation coefficient of the random variable X and Y is defined as $\rho_{XY} = \dfrac{\text{Cov}(X, Y)}{\sigma_X \sigma_Y}$.

Some of the properties of correlation coefficient are enumerated below:

1. If X and Y are random variables then,

$$|\rho_{X,Y}| \leq 1 \text{ or } -1 \leq \rho_{X,Y} \leq 1.$$

2. If $\rho_{X,Y} = \pm 1$, then $\exists\, a_1, a_2 \in \mathbb{R}, a_1 \neq 0, a_2 \neq 0$, such that,

$$Y = a_1 X + a_2.$$

Further, $\rho_{X,Y} = -1$ and $+1$ gives $a_1 < 0$ and $a_1 > 0$, respectively.

3. If X and Y are independent random variables, then,

$$\rho_{X,Y} = 0.$$

4. The random variables X and Y are perfectly negatively correlated/perfectly positively correlated/uncorrelated of $\rho_{X,Y} = -1/\rho_{X,Y} = +1/\rho_{X,Y} = 0$.

Example 2.1.11 Consider a stock whose current price (at time $t = 0$) is 100. At time $t = 1$, the stock price is a random variable, which can take the values 90, 95, 100, and 105 with probabilities of $0.1, 0.2, 0.3$, and 0.4, respectively. Determine the expected value of the stock at time $t = 1$.

Here the random variable X (stock price at time $t = 1$) takes the values $x_1 = 90$, $x_2 = 95$, $x_3 = 100$, and $x_4 = 105$. The corresponding probabilities are $p_1 = 0.1$, $p_2 = 0.2$, $p_3 = 0.3$, and $p_4 = 0.4$. Hence,

$$E(X) = \sum_{i=1}^{4} x_i\, p_i = 90 \times 0.1 + 95 \times 0.2 + 100 \times 0.3 + 105 \times 0.4 = 100.$$

Example 2.1.12 Let X and Y be two independent random variables with $\mathrm{Var}(X) = 8$, $\mathrm{Var}(Y) = 11$. Now, let $Z = 3X + 2Y$. Determine the standard deviation of the random variable Z.

Here,

$$\mathrm{Var}(Z) = \mathrm{Var}(3X + 2Y) = 9\mathrm{Var}(X) + 4\mathrm{Var}(Y) = 72 + 44 = 116.$$

Therefore, $SD(Z) = \sqrt{116} = 10.7703$.

2.2 General Probability Space

We now move on to the discussion on general probability space.

Definition 2.2.1 (σ-algebra). Let S be the non-empty sample space. A non-empty collection $\mathcal{A}$ of subsets of S is called a σ-algebra, if $\mathcal{A}$ satisfies the following properties:

(i) $S \in \mathcal{A}$.
(ii) If $A \in \mathcal{A}$, then $A^{\mathsf{c}} \in \mathcal{A}$.
(iii) For a sequence $A_1, A_2, \cdots \in \mathcal{A}$, $\bigcup_{i=1}^{\infty} A_i \in \mathcal{A}$

We are now in a position to define a measurable space and a probability space.

Definition 2.2.2 (*Measurable Space*). Let $S \neq \phi$ be a sample space with the σ-algebra $\mathcal{A}$ of subsets of S.

Then the ordered pair $(S, \mathcal{A})$ is called a "measurable space".

Definition 2.2.3 (*Probability Space*). Let $S \neq \phi$ be a sample space with the σ-algebra $\mathcal{A}$ of subsets of S and $\mathbb{P}$ be a real-valued function (probability measure) defined on $\mathcal{A}$, which satisfies the following properties:

(i) For all $A \in \mathcal{A}, 0 \leq \mathbb{P}(A) \leq 1$.
(ii) For the sample space S, $\mathbb{P}(S) = 1$.
(iii) If the events $A_i \in \mathcal{A}$, $i = 1, 2, \ldots$ are mutually exclusive, then $\mathbb{P}\left(\bigcup_{i=1}^{\infty} A_i\right) = \sum_{i=1}^{\infty} \mathbb{P}(A_i)$.

Then the ordered triplet $(S, \mathcal{A}, \mathbb{P})$ is called a "probability space".

Definition 2.2.4 (*Random Variable*). Let $(S, \mathcal{A})$ be a measurable space. A random variable $X : S \to \mathbb{R}$ is a real-valued function such that $X^{-1}(-\infty, a] = \{s : X(s) \leq a\} \in \mathcal{A}$. The random variable is also referred to as a measurable function on $(S, \mathcal{A})$.

Definition 2.2.5 (*Probability Distribution Function*). The probability distribution function of the random variable X is the function $F_X : \mathbb{R} \to [0, 1]$, defined by $F_X(s) = F_X((-\infty, s]) = \mathbb{P}_X(X \leq s)$, where F_X is non-decreasing, right-continuous, and satisfies $\lim_{s \to -\infty} F_X(s) = 0$ and $\lim_{s \to \infty} F_X(s) = 1$.

Definition 2.2.6 (*Probability Density Function*). Let X be a continuous random variable with probability distribution function F_X. Further, let there exists a non-negative integrable real-valued function $f : \mathbb{R} \to [0, \infty)$ such that

$$F_X(s) = \int_{-\infty}^{s} f(x)dx.$$

Then the function f is called the probability density function of X.

Definition 2.2.7 (*Independence of Random Variable*). The random variables $X_1, X_2, \ldots, X_n$ are said to be independent if $\mathbb{P}(X_1 \leq s_1, X_2 \leq s_2, \ldots, X_n \leq s_n) = \prod_{i=1}^{n} \mathbb{P}(X_i \leq s_i)$.

We now define the first and second moments in a general probability space.

Definition 2.2.8 (*Expectation*). Let X be a continuous random variable with the probability density function f. Then the expectation (or expected value) of the random variable X is defined as $\mu_X = E(X) = \int_{-\infty}^{\infty} x f(x) dx$, provided $\int_{-\infty}^{\infty} |x| f(x) dx$ exists. If X is a discrete random variable which takes the values $\{x_1, x_2, \dots\}$, then $E(X) = \sum_{i=1}^{\infty} x_i f(x_i)$.

Suppose that X is a continuous random variable defined on the probability space $(S, \mathcal{A}, \mathbb{P})$ and $g : \mathbb{R} \to \mathbb{R}$ is a measurable function. Then the expectation of the random variable $g(X)$ is given by $E(g(X)) = \int_{-\infty}^{\infty} g(x) f(x) dx$, provided $\int_{-\infty}^{\infty} |g(x)| f(x) dx$ exists. If X is a discrete random variable, which takes the values $\{x_1, x_2, \dots\}$, then $E[g(X)] = \sum_{i=1}^{\infty} g(x_i) f(x_i)$.

Definition 2.2.9 (*Variance, Standard Deviation, and Covariance*). Let X and Y be continuous random variables on the probability space $(S, \mathcal{A}, \mathbb{P})$. Then the variance of X is defined as,

$$\sigma_X^2 = \text{Var}(X) = E\left[(X - \mu_X)^2\right].$$

Further, the standard deviation of the random variable X is defined as,

$$\sigma_X = SD(X) = \sqrt{\text{Var}(X)}.$$

Finally, the covariance of the random variables X and Y is defined as,

$$\text{Cov}(X, Y) = E\left[(X - \mu_X)(Y - \mu_Y)\right] = E(XY) - E(X)E(Y).$$

Example 2.2.10 Let X be a continuous random variable with the probability density function being given by,

$$f(x) = \begin{cases} k(2x - x^2), & 0 < x < 1, \\ 0, & \text{otherwise.} \end{cases}$$

Determine the value of the constant k. Hence determine $P(X > 0.5)$.

Since $\int_{-\infty}^{\infty} f(x) dx = 1$, therefore,

$$\int_0^1 k(2x - x^2) dx = 1 \Rightarrow k\left[x^2 - \frac{x^3}{3}\right]_0^1 = 1 \Rightarrow k = \frac{3}{2}.$$

Therefore,

$$P(X > 0.5) = \int_{0.5}^{\infty} f(x) dx = \frac{3}{2} \int_{0.5}^1 (2x - x^2) dx = \frac{11}{16}.$$

Example 2.2.11 Let X be a continuous random variable with the probability density function being given by,

$$f(x) = \begin{cases} \frac{3}{2}(2x - x^2), & 0 < x < 1, \\ 0, & \text{otherwise.} \end{cases}$$

Determine the variance of X.

Here,

$$E(X) = \int_{-\infty}^{\infty} xf(x)\mathrm{d}x = \frac{3}{2}\int_{0}^{1} x(2x - x^2)\mathrm{d}x = \frac{3}{2}\left[\frac{2}{3}x^3 - \frac{x^4}{4}\right]_0^1 = \frac{5}{8}.$$

$$E(X^2) = \int_{-\infty}^{\infty} x^2 f(x)\mathrm{d}x = \frac{3}{2}\int_{0}^{1} x^2(2x - x^2)\mathrm{d}x = \frac{3}{2}\left[\frac{2}{4}x^4 - \frac{x^5}{5}\right]_0^1 = \frac{9}{20}.$$

Therefore,

$$\mathrm{Var}(X) = E(X^2) - (E(X))^2 = \frac{9}{20} - \frac{25}{64} = 0.059375.$$

2.3 Two Important Distributions

2.3.1 The Binomial Distribution

We consider an experiment, where there are only two possible outcomes, namely "success" or "failure", with the probability of success being p and that of failure being $1 - p$. Further, we consider the scenario where the experiment is repeated n times, and the outcomes of this experiment are independent and identically distributed. Then this experiment is known as the binomial experiment, with parameters n and p. We then consider the random variable X to denote the number of successes in the binomial experiment with parameters n and p. Now, for $n = 1$, the probability mass function of X is given by $\mathbb{P}(X = 0) = 1 - p$ and $\mathbb{P}(X = 1) = p$. Then the probability of k success is given by $\mathbb{P}(X = k) = \binom{n}{k}p^k(1 - p)^{n-k}$.

Definition 2.3.1 (*Binomial Distribution*). Let n be a positive integer and $0 < p < 1$ and let the sample space be $\mathcal{S} = \{0, 1, 2, \ldots, n\}$. Then the binomial distribution is a probability distribution on the sample space $\mathcal{S}$, with the probability mass function

$$B(k; n, p) = \binom{n}{k}p^k(1 - p)^{n-k}, \ k = 0, 1, 2, \ldots, n.$$

This gives the probability of k success in n trials of the binomial experiment with the probability of success and failure, in each trial being, p and $(1 - p)$, respectively. Note that if $X \sim B(k; n, p)$, then $E(X) = np$ and $\text{Var}(X) = np(1 - p)$. Also

$$\mathbb{P}(X \leq k) = \sum_{i=0}^{k} \binom{n}{i} p^i (1 - p)^{n-i}, \text{ and } \mathbb{P}(X = k + 1) = \frac{p}{1 - p} \frac{n - k}{k + 1} \mathbb{P}(X = k).$$

2.3.2 The Normal Distribution

If we were to consider a sample such as the height of individuals or marks earned by students in a test and plot the histogram for the dataset, we observe that it turns out to be bell-shaped. In fact, several real-life examples follow the pattern, with such random variables exhibiting a symmetric bell-shaped curve, which motivates the setup for what is known as the normal distribution.

Definition 2.3.2 (*Normal Distribution*). Let μ be a real number, and let σ be a positive real number. Then the normal distribution is a distribution whose probability density function is given by,

$$N_{\mu,\sigma}(x) = \frac{1}{\sqrt{2\pi\sigma^2}} e^{-\frac{(x-\mu)^2}{2\sigma^2}}.$$

Note that if $X \sim N_{\mu,\sigma}$, then $E(X) = \mu$ and $\text{Var}(X) = \sigma^2$. In case the mean and variance are 0 and 1, respectively, then the distribution thus obtained is called the standard normal distribution with the probability density function being $N_{0,1}(x) = \frac{1}{\sqrt{2\pi}} e^{-\frac{x^2}{2}}$. Finally, $N_{\mu,\sigma} = \mu + \sigma N_{0,1}$, i.e., if $X \sim N_{\mu,\sigma}$, then $Z = \frac{X - \mu}{\sigma} \sim N_{0,1}$.

Example 2.3.3 If four unbiased coins are tossed, then determine the probability of getting at least two tails.

The probability of getting k tails in 4 tosses of the unbiased toss (getting a tail is a success with $p = \frac{1}{2}$) is,

$$\binom{4}{k} \left(\frac{1}{2}\right)^k \left(\frac{1}{2}\right)^{4-k} = \binom{4}{k} \left(\frac{1}{2}\right)^4,$$

for $k = 0, 1, 2, 3, 4$. Then the probability of getting at least two tails is,

$$P(k \geq 2) = 1 - P(k = 0) - P(k = 1) = 1 - \binom{4}{0}\frac{1}{16} - \binom{4}{1}\frac{1}{16} = \frac{11}{16}.$$

Example 2.3.4 If $X \sim N_{\mu,\sigma}$, then prove that

$$P(\alpha < X < \beta) = \mathcal{N}_{0,1}\left(\frac{\beta - \mu}{\sigma}\right) - \mathcal{N}_{0,1}\left(\frac{\alpha - \mu}{\sigma}\right),$$

where $\mathcal{N}$ is the cumulative normal distribution function.

Let k be a constant. Then,

$$P(X < k) = P\left(\frac{X - \mu}{\sigma} < \frac{k - \mu}{\sigma}\right) = P\left(Z < \frac{k - \mu}{\sigma}\right) = \mathcal{N}_{0,1}\left(\frac{k - \mu}{\sigma}\right).$$

Therefore,

$$P(\alpha < X < \beta) = P(X < \beta) - P(X < \alpha) = \mathcal{N}_{0,1}\left(\frac{\beta - \mu}{\sigma}\right) - \mathcal{N}_{0,1}\left(\frac{\alpha - \mu}{\sigma}\right).$$

Example 2.3.5 If $X \sim B(k; n, p)$. Then prove that $E(X) = np$ and $\mathrm{Var}(X) = np(1 - p)$.

Here,

$$E(X) = \sum_{k=0}^{n} k \binom{n}{k} p^k (1 - p)^{n-k} = \sum_{k=0}^{n} n \binom{n-1}{k-1} p^k (1 - p)^{n-k}$$

$$= np \sum_{k=1}^{n} \binom{n-1}{k-1} p^{k-1} (1 - p)^{n-k} = np.$$

Further,

$$E(X^2) = \sum_{k=0}^{n} k^2 \binom{n}{k} p^k (1 - p)^{n-k} = np \sum_{k=0}^{n} k \binom{n-1}{k-1} p^{k-1}(1 - p)^{n-k},$$

$$= np \sum_{k=0}^{n} (k - 1 + 1)\binom{n-1}{k-1} p^{k-1}(1 - p)^{n-k},$$

$$= np \left(\sum_{k=0}^{n}(k - 1)\binom{n-1}{k-1} p^{k-1}(1 - p)^{n-k} + \sum_{k=0}^{n} \binom{n-1}{k-1} p^{k-1}(1 - p)^{n-k} \right),$$

$$= np \left(\sum_{k=2}^{n}(n - 1)p\binom{n-2}{k-2} p^{k-2}(1 - p)^{n-k} + \sum_{k=1}^{n} \binom{n-1}{k-1} p^{k-1}(1 - p)^{n-k} \right),$$

$$= np((n - 1)p + 1).$$

Therefore,

$$\mathrm{Var}(X) = E(X^2) - (E(X))^2 = np(np - p + 1) - (np)^2 = np(1 - p).$$

2.4 Some Important Results

In this section, we discuss a few important results, from probability, some of which we will need later,

Result 2.4.1 (Markov's Inequality). *Let X be a random variable which takes only non-negative values. Then,*

$$\mathbb{P}(X \geq a) \leq \frac{E(X)}{a} \text{ for } a > 0.$$

Result 2.4.2 (Chebyshev's Inequalities). *Let X be a random variable with the mean and variance being μ and σ^2, respectively. Then,*

$$\mathbb{P}(|X - \mu| \geq a) \leq \frac{\sigma^2}{a^2} \text{ for } a > 0.$$

Equivalently,

$$\mathbb{P}(|X - \mu| \leq a) \geq 1 - \frac{\sigma^2}{a^2} \text{ for } a > 0.$$

Result 2.4.3 (The Weak Law of Large Number). *Let $X_1, X_2, X_3, \ldots$ be a sequence of independent and identically distributed (i.i.d) random variables. Let the identical mean for each of these random variables be $E(X_i) = \mu$. Then, given $\epsilon > 0$,*

$$\mathbb{P}\left[\left|\frac{X_1 + X_2 + \cdots + X_n}{n} - \mu\right| > \epsilon\right] \to 0 \text{ as } n \to \infty$$

Result 2.4.4 (The Central Limit Theorem). *Let $X_1, X_2, X_3, \ldots$ be a sequence of independent and identically distributed (i.i.d) random variables. Let the identical mean and variance for each of these random variables be $E(X_i) = \mu$ and $\text{Var}(X_i) = \sigma^2$. Further let $S_n = \sum_{i=1}^{n} X_i$. Then the distribution of S_n for large n is approximately normally distributed with $E(S_n) = n\mu$ and $\text{Var}(S_n) = n\sigma^2$ (as $n \to \infty$). It follows that $\mathbb{P}\left[\frac{S_n - n\mu}{\sigma\sqrt{n}} < x\right] \approx \mathbb{P}(Z < x)$, where $Z \sim N_{0,1}$ (as $n \to \infty$).*

Example 2.4.5 A life insurance company has 50,000 policy holders. Let X denote the random variable for the yearly claim of policy holders, with mean of 520 and standard deviation of 820. Determine the probability that the total yearly claim will exceed 20 million.

Let X_i be the yearly claim for the policy holder i. Then the total yearly claim for the insurance company is $X = \sum_{i=1}^{50,000} X_i$. Obviously, the X_is are i.i.d. Therefore using the Central Limit Theorem, $X = \sum_{i=1}^{50,000} X_i$ is approximately normally distributed with mean of $520 \times 50{,}000 = 2.6 \times 10^7$ and standard deviation of $820\sqrt{50{,}000} = 1.83357 \times 10^5$. Therefore,

$$\mathbb{P}\left(X > 2 \times 10^7\right) = \mathbb{P}\left(\frac{X - 2.6 \times 10^7}{1.83357 \times 10^5} > \frac{2 \times 10^7 - 2.6 \times 10^7}{1.83357 \times 10^5}\right)$$

$$= \mathcal{N}_{0,1}\left(\frac{0.6 \times 10^2}{1.83357}\right) = \mathcal{N}_{0,1}(32.723) \approx 1.$$

2.5 Least Squares Estimation

Suppose that X is an input random variable, which takes the values $x_1, x_2, \ldots, x_n$ and that Y is a output random variable which takes the values $y_1, y_2, \ldots, y_n$, with y_i corresponding to x_i, for $i = 1, 2, \ldots, n$. We then consider a simple linear regression model (to fit the data points (x_i, y_i)) to be of the form,

$$Y = \alpha + \beta X + \epsilon,$$

where α, β are constants and ϵ is a random error with mean 0. Now, the actual response to the input variable x_i is the output variable y_i for $i = 1, 2, \ldots, n$. Accordingly, the error resulting from the difference between the actual value and the value given by the linear regression model is $\epsilon_i = y_i - \alpha - \beta x_i$. This gives the sum of the squares of this difference as,

$$SE^2 = \sum_{i=1}^{n} (y_i - \alpha - \beta x_i)^2.$$

In order to obtain the best model, i.e., the best estimate of α and β, in terms of x_i and y_i, we minimize the expected value of SE^2. Differentiating $E(SE^2)$, with respect to α and β, and setting equal to zero, we obtain

$$\frac{\partial E(SE^2)}{\partial \alpha} = -2\sum_{i=1}^{n}(y_i - \alpha - \beta x_i) = 0 \Rightarrow \sum_{i=1}^{n} y_i = \alpha n + \beta \sum_{i=1}^{n} x_i \Rightarrow E(Y) = \alpha + \beta E(X),$$

and

$$\frac{\partial E(SE^2)}{\partial \beta} = -2\sum_{i=1}^{n} x_i(y_i - \alpha - \beta x_i) = 0 \Rightarrow \sum_{i=1}^{n} x_i y_i = \alpha \sum_{i=1}^{n} x_i + \beta \sum_{i=1}^{n} x_i^2$$

$$\Rightarrow E(XY) = \alpha E(X) + \beta E(X^2).$$

Solving these two equations, we get $\beta = \dfrac{\mathrm{Cov}(X, Y)}{\mathrm{Var}(X)}$ and $\alpha = E(Y) - \beta E(X)$. Hence the best linear predictor is given by,

$$Y = E(Y) - \left[\frac{\mathrm{Cov}(X, Y)}{\mathrm{Var}(X)}\right] E(X) + \left[\frac{\mathrm{Cov}(X, Y)}{\mathrm{Var}(X)}\right] X + \epsilon.$$

Example 2.5.1 Let X and Y be random variables with the following tabulated values

X	Y	Probability
2	18	0.2
3	22	0.2
7	30	0.2
5	26	0.2
8	24	0.2

If we approximate Y using $\beta X + \alpha$, then determine the values of β and α.

Here $E(X) = 5$, $E(Y) = 24$. Also $\text{Cov}(X, Y) = 6.8$ and $\text{Var}(X) = 5.2$. Therefore,

$$\beta = \frac{\text{Cov}(X, Y)}{\text{Var}(X)} = \frac{6.8}{5.2} = 1.3077.$$

Finally,

$$\alpha = E(Y) - \beta E(X) = 24 - 1.3077 \times 5 = 17.4615.$$

2.6 Exercise

Exercise 2.1 Let X and Y be random variables on the finite probability space $(\mathcal{S}, \mathbb{P})$. If $E(X) = 2$ and $E(Y) = 3$, then determine $\dfrac{E(3X + 2Y)}{E(2X + 3Y)}$. Further if X and Y are independent, then determine $E(2X \times 3Y)$.

Solution:

$$\frac{E(3X + 2Y)}{E(2X + 3Y)} = \frac{3E(X) + 2E(Y)}{2E(X) + 3E(Y)} = \frac{12}{13}.$$

If X and Y are independent, then,

$$E(2X \times 3Y) = 6E(XY) = 6E(X)E(Y) = 36.$$

Exercise 2.2 If X and Y are independent random variable with the corresponding values as tabulated below, then determine $\text{Var}(3X + 2Y)$.

X	1	2	3	4	5
Y	4	6	3	2	1
Probability	$\frac{1}{5}$	$\frac{1}{5}$	$\frac{1}{5}$	$\frac{1}{5}$	$\frac{1}{5}$

Solution: Here,

$$E(X) = \frac{1}{5}(1 + 2 + 3 + 4 + 5) = 3$$

$$E(Y) = \frac{1}{5}(4 + 6 + 3 + 2 + 1) = \frac{16}{5} = 3.2.$$

X	Y	Probability	$X - E(X)$	$Y - E(Y)$
1	4	$\frac{1}{5}$	-2	0.8
2	6	$\frac{1}{5}$	-1	2.8
3	3	$\frac{1}{5}$	0	-0.2
4	2	$\frac{1}{5}$	1	-1.2
5	1	$\frac{1}{5}$	2	-2.2

Therefore,

$$\text{Var}(X) = E[(X - E(X))^2] = \frac{1}{5}(4 + 1 + 0 + 1 + 4) = 2,$$

$$\text{Var}(Y) = E[(Y - E(Y))^2] = \frac{1}{5}(0.8^2 + 2.8^2 + 0.2^2 + (-1.2)^2 + (-2.2)^2) = 2.96,$$

$$\text{Cov}(X, Y) = E[(X - E(X))(Y - E(Y))] = \frac{1}{5}(-1.6 - 2.8 + 0 - 1.2 - 4.4) = -2.$$

Hence,

$$\text{Var}(3X + 2Y) = 9\text{Var}(X) + 4\text{Var}(Y) + 12\text{Cov}(X, Y) = 5.84.$$

Exercise 2.3 Let X be a continuous random variable with the probability density function being given by,

$$f(x) = \begin{cases} ke^{-3x}, & x > 0, \\ 0, & \text{otherwise} \end{cases}.$$

Determine the value of k. Hence determine $P(X > 3)$.

Solution: Since $\int_{-\infty}^{\infty} f(x)\mathrm{d}x = 1$, therefore,

$$\int_{0}^{\infty} ke^{-3x}\mathrm{d}x = 1 \Rightarrow k\left[-\frac{e^{-3x}}{3}\right]_{0}^{\infty} = 1 \Rightarrow k = 3.$$

Therefore,

$$P(X > 3) = \int_{3}^{\infty} f(x)\mathrm{d}x = 3\int_{3}^{\infty} e^{-3x}\mathrm{d}x = \left[-e^{-3x}\right]_{3}^{\infty} = e^{-9}.$$

Exercise 2.4 Let X be a continuous random variable with probability distribution function being given by,

$$F_X(x) = \begin{cases} 0, & x < 0, \\ \frac{x}{3}, & 0 \le x \le 3, \\ 1, & x > 3. \end{cases}$$

Determine

  (i) $P(0 \leq X \leq 2)$
 (ii) $P(2 \leq X \leq 3)$
(iii) $P(Y \leq X)$, where $Y = X^3$

Solution:

  (i) $P(0 \leq X \leq 2) = F_X(2) - F_X(0) = \dfrac{2}{3} - 0 = \dfrac{2}{3}.$

 (ii) $P(2 \leq X \leq 3) = F_X(3) - F_X(2) = 1 - \dfrac{2}{3} = \dfrac{1}{3}.$

(iii) $P(Y \leq X) = P(X^3 \leq X) = P(X(X-1)(X+1) \leq 0) = P(0 \leq X \leq 1) + P(X \leq -1) = F_X(1) - F_X(0) + F_X(-1) = \frac{1}{3} - 0 + 0 = \frac{1}{3}.$

Exercise 2.5 If $X \sim B(k; 5, 0.6)$, then determine $P(X = k)$ for $k = 0, 1, 2, 3, 4, 5$.

Solution: Firstly,

$$P(X = 0) = \binom{5}{0}(0.6)^0(1 - 0.6)^{5-0} = (0.4)^5 = 0.01024.$$

Now using the recursive relation, $\mathbb{P}(X = k + 1) = \dfrac{p}{1 - p}\dfrac{n - k}{k + 1}\mathbb{P}(X = k)$, we have,

$$\mathbb{P}(X = 1) = \frac{0.6}{1 - 0.6}\frac{5 - 0}{0 + 1}P(X = 0) = 0.0768,$$

$$\mathbb{P}(X = 2) = \frac{0.6}{1 - 0.6}\frac{5 - 1}{1 + 1}P(X = 1) = 0.2304,$$

$$\mathbb{P}(X = 3) = \frac{0.6}{1 - 0.6}\frac{5 - 2}{2 + 1}P(X = 2) = 0.3456,$$

$$\mathbb{P}(X = 4) = \frac{0.6}{1 - 0.6}\frac{5 - 3}{3 + 1}P(X = 3) = 0.2592,$$

$$\mathbb{P}(X = 5) = \frac{0.6}{1 - 0.6}\frac{5 - 4}{4 + 1}P(X = 4) = 0.07776.$$

Exercise 2.6 If $X \sim N_{2,4}$, then determine

 (i) $\mathbb{P}(X > 1)$,
(ii) $\mathbb{P}(3 < X < 8)$.

Solution:

 (i) $\mathbb{P}(X > 1) = \mathbb{P}\left(\frac{X-2}{4} > \frac{1-2}{4}\right) = \mathbb{P}\left(Z > \frac{1-2}{4}\right) = \mathbb{P}(Z > -0.25) = 0.5987.$

(ii) $\mathbb{P}(3 < X < 8) = \mathbb{P}\left(\frac{3-2}{4} < \frac{X-2}{4} < \frac{8-2}{4}\right) = \mathbb{P}(0.25 < Z < 1.5) = 0.9332 - 0.5987 = 0.3345.$

Exercise 2.7 If the scores (out of 100) in a test are normally distributed with mean of 55 and variance of 16, then determine the probability that a student will score more than 70.

Solution: Let the random variable X denote the score of the student in the test. Then,

$$\mathbb{P}(X > 70) = \mathbb{P}\left(\frac{X - 55}{4} > \frac{70 - 55}{4}\right) = \mathbb{P}(Z > 3.75) = 0.001.$$

Exercise 2.8 Let X is random variable with mean 0.5 and variance 0.02273. Find the approximate value of $\mathbb{P}(0.2 \leq X \leq 0.8)$.

Solution: Here, $\mu = 0.5$ and $\sigma = \sqrt{0.02273} = 0.15$. Therefore,

$$\mathbb{P}(0.2 \leq X \leq 0.8) = \mathbb{P}(0.2 - 0.5 \leq X - \mu \leq 0.8 - 0.5) = \mathbb{P}(|X - \mu| \leq 0.3) \geq 1 - \frac{0.15^2}{0.3^2} = 0.75.$$

Hence by the Chebychev's Inequality, if we do not know the distribution of X, then $\mathbb{P}(0.2 \leq X \leq 0.8)$ is at least 0.75.

Exercise 2.9 The average temperature of a city in the first 10 days of a month are given in the following table:

Day	1	2	3	4	5	6	7	8	9	10
Temperature $°C$	17	19	17	20	19	19	19	16	19	24

Find the least square regression line $y = ax + b$. Use this least square regression line as the prediction model to estimate the likely temperature on the 17th day.

Solution:

Day(x_i)	Temperature $°C$ (y_i)	$x_i y_i$	x_i^2
1	17	17	1
2	19	38	4
3	17	51	9
4	20	80	16
5	19	95	25
6	19	114	36
7	19	133	49
8	16	128	64
9	19	171	81
10	24	240	100

Here,

$$\sum_{i=1}^{10} x_i = 55, \quad \sum_{i=1}^{10} y_i = 189, \quad \sum_{i=1}^{10} x_i y_i = 1067, \quad \sum_{i=1}^{10} x_i^2 = 385.$$

Then,

$$a = \frac{\sum_{i=1}^{10} x_i y_i - \frac{1}{10} \sum_{i=1}^{10} x_i \sum_{i=1}^{10} y_i}{\sum_{i=1}^{10} x_i^2 - \frac{1}{10} \left(\sum_{i=1}^{10} x_i\right)^2} = 0.33 \text{ and } b = \frac{1}{10} \left(\sum_{i=1}^{10} y_i - a \sum_{i=1}^{10} x_i\right) = 17.06.$$

Hence the regression line will be $y = 0.33x + 17.06$. Now at $x = 17$, we have $y = 0.33 \times 17 + 17.06 = 22.67$. So likely temperature on 17th day will be 22.67.

Asset Pricing Models

3

In Chap. 1, we dwelled upon the aspects of financial derivatives, with the asset (typically a risky security, such as stock) being the driver of the pricing of the derivatives. Accordingly, this explicit dependence on the random asset price movement needs to be modeled, in order to achieve the valuation of the financial derivative it is driving. This chapter will deliberate upon a discrete time model (binomial model) and a continuous time model (geometric Brownian motion (gBm) model), for asset price movement. Before proceeding onto the description of asset pricing models, we will briefly enumerate some of the basic notions about the time value of money, with an emphasis on interest rates. Accordingly, we let $S(0)$ be the investment made at time $t = 0$. If μ_s denotes the simple interest rate for a single period, then the amount accumulated at time $t = T$ is given by,

$$S(T) = S(0)(1 + T\mu_s).$$

On the other hand, if the investment of $S(0)$ earns interest, on the interest, then this introduces the concept of discrete compound interest rate, which we denote by μ_c, for a single period. Then the accumulated amount at time $t = T$ is given by,

$$S(T) = S(0)(1 + \mu_c)^T.$$

More generally, if the discrete compound interest rate is μ_c, per period, with the compounding happening m times per period, then an investment of $S(0)$ at time $t = 0$ accumulates to $S(T)$ at time $t = T$ and is given by,

$$S(T) = S(0)\left(1 + \frac{\mu_c}{m}\right)^{mT}.$$

Finally, if the interest is compounded at a continuous manner at rate μ_{c^*}, per period then,

$$S(T) = \lim_{m \to \infty} S(0)\left(1 + \frac{\mu_{c^*}}{m}\right)^{mT} = S(0) \lim_{m \to \infty}\left[\left(1 + \frac{\mu_{c^*}}{m}\right)^{\frac{m}{\mu_{c^*}}}\right]^{\mu_{c^*} T} = S(0)e^{\mu_{c^*} T}.$$

Here, μ_{c*} is called the continuously compounded interest rate. Until this point, we have considered the investment period to be the same as one of the time points. Suppose that an investor invests for a time period T, and in addition a fraction of a single time period, which we denote by T^* i.e., for the period $\overline{T} = T + T^*$. Then an investment of $S(0)$ at time $t = 0$ grows to an amount of $S(\overline{T})$ at time $\overline{T}$, which in the three cases become,

(i) Simple Interest:

$$S(\overline{T}) = S(0)(1 + \mu_s T) + S(0)\mu_s T^* = S(0)(1 + \mu_s \overline{T}).$$

(ii) Discrete Compound Interest:

$$S(\overline{T}) = S(0)(1 + \mu_c)^T (1 + \mu_c T^*).$$

(iii) Continuously Compound Interest:

$$S(\overline{T}) = S(0)e^{\mu_{c*} T} e^{\mu_{c*} T^*} = e^{\mu_{c*} \overline{T}}.$$

Finally, we briefly mention the concept of present value and future value. Let P_1 and P_2 be amounts at time t_1 and t_2, respectively, with $t_1 < t_2$ $(t_2 - t_1$ periods). If μ_s, μ_c, and μ_{c*} are the simple, discrete compound, and continuous compound rate per period, then P_2 is future value of P_1, and P_1 is the present value of P_2, if

$$P_1 = \frac{P_2}{(1 + \mu_s(t_2 - t_1))}, \ P_1 = \frac{P_2}{\left(1 + \frac{\mu_c}{m}\right)^{m(t_2 - t_1)}}, \text{ and } P_1 = \frac{P_2}{e^{\mu_{c*}(t_2 - t_1)}}, \text{respectively.}$$

Example 3.0.1 *An investor invests amounts of* 1000, 1250, *and* 1500 *at times* $t = 0$, $t = 1$, *and* $t = 2$, *respectively, at a continuously compounded interest rate of* 4%. *Then determine the value of the investment at time* $t = 3$.

The value of the investments made at $t = 0$, $t = 1$, and $t = 2$ till time $t = 3$ is given by $1000 \times e^{0.04 \times (3-0)}$, $1250 \times e^{0.04 \times (3-1)}$, and $1500 \times e^{0.04 \times (3-2)}$. Hence the required value is given by $1000e^{0.12} + 1250e^{0.08} + 1500e^{0.04} = 4042.8219$.

3.1 The Binomial Model of Asset Pricing

In this section, we present a discrete time model for asset pricing. Let $S(t)$ be the stock price at times $t_0 = 0$, Δt, $2\Delta t$, $\ldots$, $N\Delta t = t_N$, i.e., t takes the values $t_i = i\Delta t$, $i = 0, 1, \ldots, N$, with $\Delta t = \dfrac{t_N - t_0}{N}$. For the sake of brevity, we identify $S(t)$ at time $t = t_i$ as $S(i)$, $i = 0, 1, 2, \ldots, N$. Let us begin the description with a one step model for asset price $S(1)$, from $S(0)$. The binomial model gives the value of $S(1)$, as the random variable,

$$S(1) = \begin{cases} S(0)(1 + u), & \text{with probabilty } p, \\ S(0)(1 + d), & \text{with probabilty } 1 - p. \end{cases}$$

This means that the model predicts, that from time t_0 to t_1, the stock price can either move up by a factor of $(1 + u)$, or down by a factor of $(1 + d)$, with the respective probabilities of p and $1 - p$. Note that, here the terms up and down have been used in the sense of $u > d$, and that it is possible that $S(0)(1 + d)$ may be greater than $S(0)$. The return variable between t_0 and t_1, then, is given by,

$$r(1) = \begin{cases} u, & \text{with probabilty } p, \\ d, & \text{with probabilty } (1 - p). \end{cases}$$

In general, the return variable between t_{i-1} and t_i is given by,

$$r(i) = \begin{cases} u, & \text{with probabilty } p, \\ d, & \text{with probabilty } (1 - p). \end{cases}$$

with $(1 + u)$ and $(1 + d)$ being the up and the down factor, respectively, between time t_{i-1} and t_i. Then the stock price at time t_N is a random variable, which takes the values,

$$S(N) = S(0)(1 + u)^k (1 + d)^{N-k},$$

with probability $\binom{N}{k} p^k (1 + d)^{N-k}$, $k = 0, 1, \ldots, N$. This is the value of $S(N)$ (starting from $S(0)$), resulting from k upward, and $N - k$ downward movement, from time t_0 to t_N. While the discrete time models are amenable in terms of the tractability of the mathematics used, they exhibit the disadvantage of limiting the number of possible values that the asset price may take and the time points at which this asset price movement takes place, which is out of sync with the continuous, as well as high frequency trading, that is the norm in today's financial markets. Accordingly, we move toward an improved continuous time asset pricing model, as a limiting case of a specifically chosen sequence of the binomial model.

Example 3.1.1 *Consider a binomial model with $S(0) = 100$, $u = 0.1$, $d = -0.1$, and $p = \frac{2}{3}$. Determine the expected value of $S(2)$.*

The values of $S(2)$ along with the respective probability are:

$S(2)$	Probability
$S(0)(1 + u)^2 = 121$	$p^2 = \dfrac{4}{9}$
$S(0)(1 + u)(1 + d) = 99$	$2p(1 - p) = \dfrac{4}{9}$
$S(0)(1 + d)^2 = 81$	$(1 - p)^2 = \dfrac{1}{9}$

Hence the required expected value is,

$$E(S(2)) = 121 \times \frac{4}{9} + 99 \times \frac{4}{9} + 81 \times \frac{1}{9} = 106.7778.$$

3.2 The gBm Model

Let us consider a bond, and a stock between times t and $t + \Delta t$, with N being the number of time steps between t_0 and t_N. Let $B^{(N)}(t)$ be the price of the bond at time t. Then its value at time $t + \Delta t$ is given by,

$$B^{(N)}(t + \Delta t) = \left(1 + \mu_f^{(N)}\right) B^{(N)}(t),$$

where $\mu_f^{(N)}$ is the riskfree rate. Further, let $S^{(N)}(t)$ be the price of the stock at time t. Then its value at time $t + \Delta t$ is given by,

$$S^{(N)}(t + \Delta t) = (1 + r^{(N)}(t))S^{(N)}(t),$$

where

$$r^{(N)}(t) = \begin{cases} u^{(N)}, & \text{with probabilty } p, \\ d^{(N)}, & \text{with probabilty } (1-p), \end{cases}$$

with $1 + u^{(N)}$ and $1 + d^{(N)}$ being the up and the down factor, respectively. Note that this entire setup is for a fixed N, and $d^{(N)} < \mu_f^{(N)} < u^{(N)}$. Now we consider a particular case of $p = \dfrac{1}{2} = (1 - p)$ (the proof without this condition is more involved, and beyond the scope of this book). Now, while moving to continuous time limit (with $\Delta t \to 0$, i.e., $N \to \infty$), we get for the bond,

$$B(t) = e^{r_c * t} B(0), \text{ with } 1 + \mu_f^{(N)} = e^{r_c * \Delta t}.$$

Also, the log return of the random variable $r^{(N)}(t)$ is given by,

$$r_l^{(N)}(t) = \ln\left(1 + r^{(N)}(t)\right) = \ln\left(\frac{S^{(N)}(t + \Delta t)}{S^{(N)}(t)}\right),$$

for, $t = t_i = i\Delta t, i = 0, 1, \ldots, N - 1$. Further,

$$r_l^{(N)}(t) = \begin{cases} \ln\left(1 + u^{(N)}\right), & \text{with probability } p, \\ \ln\left(1 + d^{(N)}\right), & \text{with probability } (1 - p). \end{cases}$$

Since $r_l^{(N)}(t)$ are independent and identically distributed, therefore for $t = i\Delta t$,

$$r_l^{(N)}(\Delta t) = r_l^{(N)}(2\Delta t) = \cdots = r_l^{(N)}((i-1)\Delta t) = r_l^{(N)}(i\Delta t).$$

$$\therefore r_l^{(N)}(i\Delta t) + r_l^{(N)}((i-1)\Delta t) + \cdots + r_l^{(N)}(2\Delta t) + r_l^{(N)}(\Delta t) = r_l^{(N)}(0, i\Delta t),$$

where

$$r_l^{(N)}(j\Delta t, i\Delta t) = \ln\left(\frac{S^{(N)}(i\Delta t)}{S^{(N)}(j\Delta t)}\right).$$

$$\therefore E(r_l^{(N)}(0, i\Delta t)) = i E(r_l^{(N)}(\Delta t)),$$

and

$$\text{Var}(r_l^{(N)}(0, i\Delta t)) = i\,Var(r_l^{(N)}(\Delta t)),$$

using the i.i.d. property.

Now suppose $E\left(r_l^{(N)}(0, i\Delta t)\right) = \mu t = \mu(i\Delta t)$ and $\text{Var}\left(r_l^{(N)}(0, i\Delta t)\right) = \sigma^2 t = \sigma^2(i\Delta t)$, for some $\mu \in \mathbb{R}$ and $\sigma > 0$.
Then,

$$i E\left(r_l^{(N)}(\Delta t)\right) = \mu(i\Delta t) \Rightarrow \mu = \frac{1}{\Delta t} E\left(r_l^{(N)}(\Delta t)\right),$$

and

$$i\,\text{Var}\left(r_l^{(N)}(\Delta t)\right) = \sigma(i\Delta t) \Rightarrow \sigma = \frac{1}{\Delta t}\text{Var}(r_l^{(N)}(\Delta t)).$$

The values of μ and σ^2 are the mean and variance of the log return per unit time, an interpretation that will become relevant later on. Now,

$$\mu = \frac{1}{\Delta t} E(r_l^{(N)}(\Delta t)) \Rightarrow \mu\Delta t = \frac{1}{2}\left[\ln\left(1 + u^{(N)}\right) + \ln\left(1 + d^{(N)}\right)\right],$$

and,

$$\sigma = \frac{1}{\Delta t}\left(Var(r_l^{(N)}(\Delta t))\right),$$

$$\Rightarrow \sigma^2\Delta t = \frac{1}{4}\left[\ln(1 + u^{(N)}) - \ln(1 + d^{(N)})\right]^2,$$

$$\Rightarrow \sigma\sqrt{\Delta t} = \frac{1}{2}\left(\ln\left[1 + u^{(N)}\right] - \ln\left[1 + d^{(N)}\right]\right).$$

Solving we get,

$$1 + u^{(N)} = e^{\mu\Delta t + \sigma\sqrt{\Delta t}},$$

and

$$1 + d^{(N)} = e^{\mu\Delta t - \sigma\sqrt{\Delta t}}.$$

We define a new random variable X_{b_k} which takes the values $+1$ and -1 (with equal probability) for $k = 1, 2, \ldots, N$. Then,

$$S^{(N)}(\Delta t) = S(0)e^{\mu\Delta t + X_{b_1}\sigma\sqrt{\Delta t}}.$$

Now,

$$S^{(N)}(i\Delta t) = S(0)e^{\mu i\Delta t + \sum_{k=1}^{i} X_{b_k}\sigma\sqrt{\Delta t}} = S(0)e^{\mu t + \sigma W^{(N)}(t)},$$

where

$$W^{(N)}(t) = W^{(N)}(i\Delta t) = \sqrt{\Delta t}\left(\sum_{k=1}^{i} X_{b_k}\right).$$

Observe that $W^{(N)}(0) = 0$. For the time window $[0, t_N]$,

$$W^{(N)}(t_N) = \sqrt{\Delta t}\left(\sum_{k=1}^{N} X_{b_k}\right) \Rightarrow \frac{W^{(N)}(t_N)}{\sqrt{t_N}} = \frac{\sum_{k=1}^{N} X_{b_k}}{\sqrt{N}}.$$

As $N \to \infty$, $\Delta t \to 0$, then using the Central Limit Theorem, we obtain,

$$\sum_{k=1}^{N} \frac{X_{b_k}}{\sqrt{N}} \to N_{0,1} \text{ as } N \to \infty.$$

This motivates the definition of a Wiener process which is a stochastic process (or a collection of random variables $W(t)$), with $W(0) = 0$, $W(t) - W(s) \sim N_{0,\sqrt{t-s}}$, $0 \leq s < t$ distribution, and $W(t_N) - W(t_{N-1})$, $W(t_{N-1}) - W(t_{N-2})$, ..., $W(t_1) - W(t_0)$ are independent. Coming back to,

$$S^{(N)}(t) = S(0)e^{\mu t + \sigma W^{(N)}(t)},$$

as $N \to \infty$, this relation reduces to,

$$S(t) = S(0)e^{\mu t + \sigma W(t)}.$$

It can be shown that, this is equivalent to the SDE,

$$dS(t) = \left(\mu + \frac{1}{2}\sigma^2\right) S(t)dt + \sigma S(t)dW(t).$$

If this SDE is of the form,

$$dS(t) = \nu S(t)dt + \sigma S(t)dW(t),$$

then its solution is given by,

$$S(t) = S(0)e^{\left(\nu - \frac{1}{2}\sigma^2\right)t + \sigma W(t)}.$$

Example 3.2.1 *For a gBm model, $dS(t) = \mu S(t)dt + \sigma S(t)dW(t)$, prove that $E(S(t)) = S(0)e^{\mu t}$.*

For the gBm

$$S(t) = S(0)e^{\left(\mu - \frac{1}{2}\sigma^2\right)t + \sigma W(t)}.$$

Hence,

$$E[S(t)] = E\left[S(0)e^{\left(\mu - \frac{1}{2}\sigma^2\right)t + \sigma W(t)} \right]$$

$$= S(0)e^{\left(\mu - \frac{1}{2}\sigma^2\right)t} E\left[e^{\sigma W(t)} \right]$$

$$= S(0)e^{\left(\mu - \frac{1}{2}\sigma^2\right)t} \int_{-\infty}^{\infty} \frac{1}{\sqrt{2\pi t}\sigma} e^x e^{-\frac{1}{2}\left(\frac{x^2}{\sigma^2 t}\right)}$$

$$= S(0)e^{\left(\mu - \frac{1}{2}\sigma^2\right)t} \int_{-\infty}^{\infty} \frac{1}{\sqrt{2\pi t}\sigma} e^{-\frac{1}{2\sigma^2 t}(x - \sigma^2 t)^2 + \frac{1}{2}\sigma^2 t}$$

$$= S(0)e^{\left(\mu - \frac{1}{2}\sigma^2\right)t} e^{\frac{1}{2}\sigma^2 t} \int_{-\infty}^{\infty} \frac{1}{\sqrt{2\pi t}\sigma} e^{-\frac{1}{2\sigma^2 t}(x - \sigma^2 t)^2}$$

$$= S(0)e^{\mu t}$$

3.3 Exercise

Exercise 3.1 An investment of 1000 gives values of $v_{1,d}$ and $v_{1,c}$ after 1 year, using annual rate of 6% compounded monthly and annual rate of 6% compounded continuously, respectively. Determine $v_{1,c} - v_{1,d}$.

Solution: Here,

$$v_{1,d} = 1000 \left(1 + \frac{0.06}{12}\right)^{12} = 1061.67781,$$

$$v_{1,c} = 1000 e^{0.06 \times 1} = 1061.83655.$$

Hence $v_{1,c} - v_{1,d} = 0.15874$.

Exercise 3.2 Consider a binomial model with $S(0) = 100$, $u = 0.2$, $d = -0.1$, and $p = \frac{1}{2}$. If M and m denote the maximum and minimum value of $S(3)$, respectively, then determine $\frac{M}{m}$.

Solution: Since $M = S(0)(1 + u)^3$ and $m = S(0)(1 + d)^3$, therefore $\frac{M}{m} = \left(\frac{1+u}{1+d}\right)^3 = \left(\frac{1.2}{0.9}\right)^3 = 2.37$.

Exercise 3.3 Consider a binomial model with $S(0) = 100$, $u = 0.15$, $d = -0.05$, and $p = \frac{3}{4}$. If $K = 105$, then determine the expected value of the random variable, $\max(S(3) - K, 0)$.

Solution: The value of $S(3)$ and $\max(S(3) - K, 0)$ along with the respective probabilities are tabulated.

$S(3)$	$\max(S(3) - K, 0)$	Probability
$S(0)(1 + u)^3 = 152.0875$	47.0875	$p^3 = \dfrac{27}{64}$
$S(0)(1 + u)^2(1 + d) = 125.6375$	20.6375	$3p^2(1 - p) = \dfrac{27}{64}$
$S(0)(1 + u)(1 + d)^2 = 103.7875$	0	$3p(1 - p)^2 = \dfrac{9}{64}$
$S(0)(1 + d)^3 = 85.7375$	0	$(1 - p)^3 = \dfrac{1}{64}$

Hence the required expected value is,

$$E[\max(S(3) - K, 0)] = 47.0875 \times \frac{27}{64} + 20.6375 \times \frac{27}{64} + 0 + 0 = 28.5715.$$

Exercise 3.4 Determine the stochastic differential equation (SDE) corresponding to $S(t) = S(0)e^{3t+2W(t)}$.

Solution: Here, $\mu = 3$ and $\sigma = 2$. Therefore

$$dS(t) = \left(\mu + \frac{1}{2}\sigma^2\right) S(t)dt + \sigma S(t)dW(t)$$
$$= \left(3 + \frac{1}{2} \times 2^2\right) S(t)dt + 2 S(t)dW(t)$$
$$= 5S(t)dt + 2S(t)dW(t).$$

Exercise 3.5 If $S(t) := S(0)e^{2.5t+\sqrt{3}W(t)}$, then determine $E(S(3))$.

Solution: Recall that $E(S(t)) = S(0)e^{\mu t}$. Therefore $E(S(3)) = S(0)e^{2.5\times3}$ $= 1808.04 \times S(0)$.

Mean-Variance Portfolio Theory

4

Given the huge array of investment alternatives available in a market, such as basic securities and derivatives, the investors' choice needs to be made simply by taking into consideration only a limited number of such alternatives, to achieve an optimal collection of such assets or the best possible portfolio. A portfolio is simply a collection of a number of assets, and can be specified as a tuple of number of units of each asset held by the investor, in the portfolio. The portfolio management process involves the sequential steps of security analysis, portfolio analysis, and portfolio selection. In the security analysis phase of this execution, the focus is on the prediction of the returns on the securities, for a single time period into the future, known as the holding period or the investment horizon. The portfolio analysis is an approach based on three statistical inputs on the returns of the assets, being considered for the portfolio, namely the expected rate of return, the variance (or equivalently the standard deviation) of returns, and the covariance (or equivalently correlation coefficient) between the returns of all pairs of assets in the portfolio. In the last stage of the portfolio selection process, the goal is to choose the best portfolio from what is known as the efficient frontier, and is based on the expected return and risk of the portfolio, which are the two pillars of modern portfolio theory.

4.1 Return and Risk of a Portfolio

At an intuitive level, the objective from the point of view of an investor would amount to the maximization of the trade-off between the return and risk of a portfolio. More specifically, the goal would be to maximize the expected return, while at the same time, minimizing the risk. Accordingly, we begin our discussion with the mathematical formulation of return and risk of a portfolio. Let us consider a portfolio comprising of n assets. We consider a single period investment horizon, with the beginning and the end of the investment horizon being time $t = 0$ and $t = 1$, respectively. Further, if the values of the i-th asset at time $t = 0$ and time $t = 1$ are

© The Author(s), under exclusive license to Springer Nature Singapore Pte Ltd. 2023
S. P. Chakrabarty and A. Kanaujiya, *Mathematical Portfolio Theory and Analysis*,
Compact Textbooks in Mathematics, https://doi.org/10.1007/978-981-19-8544-7_4

denoted by $v_{i,0}$ and $v_{i,1}$, respectively, then the return on the single period investment for the i-th asset is given by,

$$r_i = \frac{v_{i,1} - v_{i,0}}{v_{i,0}}.$$

Now, if we consider a portfolio of n assets, with the number of the units of i-th asset being $N_i, i = 1, 2, \ldots, n$, then the amount invested in the i-th asset at time $t = 0$ is $N_i v_{i,0}$, and accordingly the total amount invested in the portfolio, at time $t = 0$ will be $\sum_{i=1}^{n} N_i v_{i,0}$. This brings us to the definition of the weight of the i-th asset in the portfolio being given by,

$$w_i = \frac{N_i v_{i,0}}{\sum_{i=1}^{n} N_i v_{i,0}}.$$

Here, the weight w_i is the proportion of the total investment, that is made in the i-th asset. Note that $N_i > 0 (\equiv w_i > 0)$ indicates a long position and $N_i < 0 (\equiv w_i < 0)$ indicates a short position. In particular, if the asset is a stock, then $w_i < 0$ indicates a short-selling position on the i-th asset. Observe that $\sum_{i=1}^{n} w_i = 1$. Now, if we make an investment in N_i units of the i-th asset at time $t = 0$ and hold on to the investment until time $t = 1$, in the portfolio, then, the value of the portfolio at time $t = 0$ and time $t = 1$ are $\sum_{i=1}^{n} N_i v_{i,0}$ and $\sum_{i=1}^{n} N_i v_{i,1}$, respectively. Therefore, the return on the portfolio (say P) during the investment horizon is given by,

$$\begin{aligned} r_P &= \frac{\sum_{i=1}^{n} N_i v_{i1} - \sum_{i=1}^{n} N_i v_{i,0}}{\sum_{i=1}^{n} N_i v_{i,0}} \\ &= \sum_{i=1}^{n} \left[\frac{N_i (v_{i,1} - v_{i,0})}{\sum_{i=1}^{n} N_i v_{i,0}} \right] \\ &= \sum_{i=1}^{n} \left[\frac{N_i r_i v_{i0}}{\sum_{i=1}^{n} N_i v_{i,0}} \right] \\ &= \sum_{i=1}^{n} \left[\frac{N_i v_{i,0}}{\sum_{i=1}^{n} N_i v_{i,0}} \times r_i \right] \\ &= \sum_{i=1}^{n} w_i r_i. \end{aligned}$$

Thus the return of the portfolio is equal to the weighted sum of the returns of the constituent assets of the portfolio. Since $v_{i,1}$ is a random variable in case of a risky asset, so the return on the portfolio, r_P is also a random variable. Taking expectation on both sides, and using the linearity property of expectation, we get the expected return on the portfolio as,

$$\mu_P = E(r_P) = E \left[\sum_{i=1}^{n} w_i r_i \right] = \sum_{i=1}^{n} w_i E(r_i) = \sum_{i=1}^{n} w_i \mu_i,$$

where $\mu_i = E(r_i)$ is the expected return on the i-th asset. Thus we get the result, that,

Result 4.1.1 (Expected Return of a Portfolio) *If a portfolio P comprises of n assets, with the return r_i and weight w_i, for the i-th asset, then the return of the portfolio P is given by,*

$$r_P = \sum_{i=1}^{n} w_i r_i.$$

Further the expected return of the portfolio P is given by,

$$\mu_P = E(r_P) = \sum_{i=1}^{n} w_i E(r_i) = \sum_{i=1}^{n} w_i \mu_i.$$

The uncertainty associated with a risky asset is intricately linked to the variability of its return from a reference point, the natural choice for which is the expected return. Further, the greater the deviation of the return (random variable) from the expected return, the more undesirable the investment choice is, and consequently, the "dispersion" of the return variable from the expected value, can be viewed as a manifestation of the risk associated with an investment in the asset. Accordingly, this motivates the choice of the variance (or equivalently the standard deviation) of return of an asset, as the measure of the risk associated with it. Thus, the variance of the return of the i-th asset is defined as,

$$\sigma_i^2 = E\,[r_i - E(r_i)]^2 = E\,[r_i - \mu_i]^2\,,$$

and its standard deviation is defined as,

$$\sigma_i = \sqrt{\sigma_i^2} = \sqrt{E\,[r_i - E(r_i)]^2} = \sqrt{E\,[r_i - \mu_i]^2}.$$

Recall that the return of a portfolio of n assets is given by,

$$r_P = \sum_{i=1}^{n} w_i r_i.$$

Taking the variance on both the sides, we obtain,

$$\sigma_P^2 = \text{Var}(r_P) = \text{Var}\left(\sum_{i=1}^{n} w_i r_i\right)$$

$$= \sum_{i=1}^{n}\sum_{j=1}^{n} w_i w_j \text{Cov}(r_i, r_j)$$

$$= \sum_{i=1}^{n}\sum_{j=1}^{n} w_i w_j \sigma_{ij}$$

$$= \sum_{i=1}^{n}\sum_{j=1}^{n} w_i w_j \rho_{ij} \sigma_i \sigma_j,$$

where $\sigma_{ij} = \text{Cov}(r_i, r_j)$ and $\rho_{ij} = \dfrac{\sigma_{ij}}{\sigma_i \sigma_j}$ are the covariance and correlation of the returns of the i-th and the j-th asset.

Result 4.1.2 (Risk of a Portfolio) *If a portfolio P comprises of n assets with return r_i and weight w_i, for the i-th asset, then the variance of return of the portfolio P is given by,*

$$\sigma_P^2 = \sum_{i=1}^{n} \sum_{j=1}^{n} w_i w_j \sigma_{ij} = \sum_{i=1}^{n} \sum_{j=1}^{n} w_i w_j \rho_{ij} \sigma_i \sigma_j.$$

Further, the standard deviation of return of the portfolio P is given by,

$$\sigma_P = \sqrt{\sigma_P^2} = \sqrt{\sum_{i=1}^{n} \sum_{j=1}^{n} w_i w_j \sigma_{ij}} = \sqrt{\sum_{i=1}^{n} \sum_{j=1}^{n} w_i w_j \rho_{ij} \sigma_i \sigma_j}.$$

Note that for a risky asset $\sigma_i > 0$.

4.2   Estimation of Expected Return, Variance and Covariance

The analysis of portfolio would require information about the weights and the estimation of expected return, variance, covariance, and correlation coefficient of the returns of the assets. This latter set of estimation can be carried out based on the historical data of return of the asset, that is, the random variable r_i for the return of the i-th asset is assumed to take the values of the historical returns with equal likelihood. If there are $K + 1$ historical asset values (in which case there would be K historical values of return, which are taken to be the values that the random variable r_i takes, with equal probability of $\frac{1}{K}$), then the expected return, variance, covariance and correlation coefficient of the returns of the i-th and j-th asset are given by,

$$\widehat{\mu}_i = \frac{1}{K} \sum_{k=1}^{K} r_{ik}, \text{ where } r_{ik} \text{ is the } k\text{-th historical return for the } i\text{-th asset,}$$

$$\widehat{\sigma}_i^2 = \frac{1}{K-1} \sum_{k=1}^{K} (r_{ik} - \widehat{\mu}_i)^2,$$

$$\widehat{\sigma}_{ij} = \frac{1}{K-1} \sum_{k=1}^{K} (r_{ik} - \widehat{\mu}_i)(r_{jk} - \widehat{\mu}_j), \text{ and,}$$

$$\widehat{\rho}_{ij} = \frac{\widehat{\sigma}_{ij}}{\widehat{\sigma}_i \widehat{\sigma}_j}.$$

Example 4.2.1 Consider a stock which is purchased for an amount of 100 and its value after one year takes the following possible values, depending on the state of the economy. Determine the expected return and variance of the returns (r_i) of the stock (in percentage).

State	Value	Probability
Recession	85	0.2
Stagnation	100	0.4
Boom	110	0.4

The return table is as given below. Therefore, $E(r_i) = 0.2 \times (-0.15) + 0.4 \times (0)$ $+ 0.4 \times 0.1 = 0.01$ or 1%. Hence $\text{Var}(r_i) = (-0.15 - 0.01)^2 \times 0.2 + (0 - 0.01)^2 \times 0.4 + (0.1 - 0.01)^2 \times 0.4 = 0.00836$ or 0.836%.

State	Value	Return	Probability
Recession	85	−0.15	0.2
Stagnation	110	0	0.4
Boom	110	0.1	0.4

Example 4.2.2 Suppose that we have a portfolio comprising of 20 stocks of asset a_1 and 20 stocks of asset a_2, with the respective current prices being $vs_1(0) = 30$ and $vs_2(0) = 70$. If the weights of a_1 and a_2 (at time $t = 0$) are w_1 and w_2, respectively, then determine the value of $\dfrac{3w_1}{w_2}$.

At time $t = 0$, $w_1 = \dfrac{20 \times 30}{20 \times 30 + 20 \times 70} = 0.3$. Hence $w_2 = 0.7$. Therefore $\dfrac{3w_1}{w_2} = \dfrac{9}{7}$.

4.3 The Mean-Variance Portfolio Analysis

We commence our discussion on the mean-variance portfolio analysis, an approach due to Harry Markowitz, which is based on certain basic assumptions, namely

(1) Investors are driven only by the mean and variance of return of assets.
(2) Investors prefer higher mean to lower mean and lower variance to higher variance.
(3) The information about mean, variance, and covariance, for all assets being considered for inclusion in the portfolio, is known.

We need two key definitions, required for this framework, namely the "opportunity set or feasible set" and the "efficient frontier".

Definition 4.3.1 *(Opportunity Set or Feasible Set)* The opportunity set or feasible set is the set of all portfolios that can be formed from n securities (including the possibility of the portfolio comprising exclusively of one of the n securities). It is also equivalently, defined as the set of all possible pairs (σ_P, μ_P), of standard deviation and expected return.

We begin the discussion of the opportunity set or feasible set with the case of a two-asset portfolio. Let us consider a portfolio P of two risky assets, with the returns being r_1 and r_2, and the respective weights being w_1 and w_2. Then the expected return and risk of the portfolio P, are given by,

$$\mu_P = w_1\mu_1 + w_2\mu_2,$$

and

$$\sigma_P^2 = w_1^2\sigma_1^2 + w_2^2\sigma_2^2 + 2w_1w_2\sigma_{12},$$

respectively. So the opportunity set is the set of points (σ_P, μ_P) as above, resulting from all the paired values (w_1, w_2), satisfying $w_1 + w_2 = 1$.

4.3.1   Minimum Variance Portfolio for Two Risky Assets

As a natural next step, we will now examine the approach to the minimization of the risk of the two-asset portfolio, i.e., minimization of σ_P^2. We first consider the case that both $\rho_{12} = 1$ and $\sigma_1 = \sigma_2$, do not hold simultaneously. Using $w_2 = 1 - w_1$, we obtain,

$$\sigma_P^2 = w_1^2\sigma_1^2 + (1 - w_1)^2\sigma_2^2 + 2w_1(1 - w_1)\sigma_{12}.$$

Taking the derivative of σ_P^2 with respect to w_1 and setting it equal to zero results in,

$$\frac{d\sigma_P^2}{dw_1} = 2w_1\sigma_1^2 - 2(1 - w_1)\sigma_2^2 + 2(1 - 2w_1)\sigma_{12} = 0$$

$$\Rightarrow w_1\left(\sigma_1^2 + \sigma_2^2 - 2\sigma_{12}\right) - \sigma_2^2 + \sigma_{12} = 0$$

$$\Rightarrow w_1 = \frac{\sigma_2^2 - \sigma_{12}}{\sigma_1^2 + \sigma_2^2 - 2\sigma_{12}}.$$

Consequently,

$$w_2 = 1 - w_1 = \frac{\sigma_1^2 - \sigma_{12}}{\sigma_1^2 + \sigma_2^2 - 2\sigma_{12}}.$$

Further,

$$\frac{d^2\sigma_P^2}{dw_1^2} = 2\sigma_1^2 + 2\sigma_2^2 - 4\sigma_{12} = 2(\sigma_1^2 + \sigma_2^2 - 2\sigma_{12}) = 2(\sigma_1^2 + \sigma_2^2 - 2\rho_{12}\sigma_1\sigma_2).$$

Since both $\rho_{12} = 1$ and $\sigma_1 = \sigma_2$ do not hold simultaneously, it can be seen that (using $\rho_{12} \leq 1$), we have $\dfrac{d^2\sigma_P^2}{dw_1^2} > 0$.

Alternatively, if $\rho_{12} = 1$ and $\sigma_1 = \sigma_2$ hold simultaneously, then,

$$\sigma_P^2 = w_1^2\sigma_1^2 + (1 - w_1)^2\sigma_2^2 + 2w_1(1 - w_1)\sigma_1\sigma_2 = [w_1\sigma_1 + (1 - w_1)\sigma_2]^2.$$

In this case, the minimum value of σ_P^2 can be zero, provided,

$$w_1\sigma_1 + (1 - w_1)\sigma_2 = 0 \Rightarrow w_1(\sigma_1 - \sigma_2) + \sigma_2 = 0 \Rightarrow w_1 = \frac{\sigma_2}{\sigma_2 - \sigma_1},$$

and consequently

$$w_2 = -\frac{\sigma_1}{\sigma_2 - \sigma_1} = \frac{\sigma_1}{\sigma_1 - \sigma_2}.$$

Theorem 4.3.2 (Minimum Variance Portfolio for Two Assets) *Consider a portfolio P, comprising of two assets with returns r_i ($i = 1, 2$), standard deviation of returns σ_i ($i = 1, 2$), covariance of returns σ_{12}, and correlation of returns ρ_{12}. Further assume that the conditions $\rho_{12} = 1$ and $\sigma_1 = \sigma_2$ do not hold simultaneously. Then the minimum variance portfolio is given by,*

$$w_1^{mvp} = \frac{\sigma_2^2 - \sigma_{12}}{\sigma_1^2 + \sigma_2^2 - 2\sigma_{12}} \text{ and } w_2^{mvp} = \frac{\sigma_1^2 - \sigma_{12}}{\sigma_1^2 + \sigma_2^2 - 2\sigma_{12}}.$$

Further, in case both the conditions $\rho_{12} = 1$ and $\sigma_1 = \sigma_2$, hold simultaneously, then the minimum variance portfolio is given by,

$$w_1^{mvp} = \frac{\sigma_2}{\sigma_2 - \sigma_1} \text{ and } w_2^{mvp} = \frac{\sigma_1}{\sigma_1 - \sigma_2}.$$

Finally, the expected return for the minimum variance portfolio is given by,

$$\mu_P^{mvp} = w_1^{mvp}\mu_1 + w_2^{mvp}\mu_2.$$

4.3.2   Minimum Variance Portfolio for n Risky Assets

We now consider a portfolio of n risky assets with the return and weights of i-th asset being r_i and w_i respectively. For the sake of brevity, we introduce the following vector and matrix notation:

1. Weight vector: $\mathbf{w} = \begin{pmatrix} w_1 & w_2 & \ldots & w_n \end{pmatrix}$.
2. Unit vector: $\mathbf{u} = \begin{pmatrix} 1 & 1 & \ldots & 1 \end{pmatrix}$.
3. Return vector: $\boldsymbol{\mu} = \begin{pmatrix} \mu_1 & \mu_2 & \ldots & \mu_n \end{pmatrix}$.
4. Covariance matrix: $C = \begin{pmatrix} \sigma_{ij} \end{pmatrix}$, $i, j = 1, 2, \ldots, n$.

Note that C is symmetric matrix (i.e., $C^\top = C$) and is positive semi-definite.
Then the expected return and variance of the portfolio P is given by $\mu_P = \boldsymbol{\mu}\mathbf{w}^\top$ and $\sigma_P^2 = \mathbf{w}C\mathbf{w}^\top$. Also the statutory condition $\sum_{i=1}^{n} w_i = 1$ is represented as $\mathbf{u}\mathbf{w}^\top = 1$. Now the problem of minimization of the variance of portfolio P reduces to determination of weight vector $\mathbf{w}$ which minimizes $\sigma_P^2 = \mathbf{w}C\mathbf{w}^\top$ subject to $\mathbf{u}\mathbf{w}^\top = 1$. This problem can no longer be dealt with the minimization approach for a function of single variable, as was done in the case of a two-asset portfolio.

Consequently, we then have to resort to the technique of Lagrange multiplier on the following problem:

$$\min \sigma_P^2 = \mathbf{w} C \mathbf{w}^\top \text{ subject to } \mathbf{u} \mathbf{w}^\top = 1.$$

Accordingly, we define the Lagrangian as,

$$F(\mathbf{w}, \lambda) = \mathbf{w} C \mathbf{w}^\top + \lambda \left(1 - \mathbf{u} \mathbf{w}^\top\right).$$

Taking the partial derivative of $F(\mathbf{w}, \lambda)$ with respect to each $w_i, i = 1, 2, \ldots, n$ as well as λ, and setting them equal to zero, we obtain,

$$C \mathbf{w}^\top = \frac{\lambda}{2} \mathbf{u}^\top \text{ and } \mathbf{u} \mathbf{w}^\top = 1.$$

Now,

$$C \mathbf{w}^\top = \frac{\lambda}{2} \mathbf{u}^\top \Rightarrow \mathbf{w} C^\top = \frac{\lambda}{2} \mathbf{u} \Rightarrow \mathbf{w} C = \frac{\lambda}{2} \mathbf{u} \Rightarrow \mathbf{w} = \frac{\lambda}{2} \mathbf{u} C^{-1} \left(\text{since } C^\top = C\right).$$

Substituting this $\mathbf{w}$ in $\mathbf{u} \mathbf{w}^\top = 1$, we get,

$$\mathbf{u} \left(\frac{\lambda}{2} \mathbf{u} C^{-1}\right)^\top = 1 \Rightarrow \frac{\lambda}{2} \mathbf{u} (C^{-1})^T \mathbf{u}^\top = 1 \Rightarrow \frac{\lambda}{2} = \frac{1}{\mathbf{u} C^{-1} \mathbf{u}^\top}.$$

Thus we have,

$$\mathbf{w} = \frac{\lambda}{2} \mathbf{u} C^{-1} = \frac{\mathbf{u} C^{-1}}{\mathbf{u} C^{-1} \mathbf{u}^\top}.$$

Theorem 4.3.3 (Minimum Variance Portfolio for n Assets) *Consider a portfolio P, comprising of n assets with the unit vector $\mathbf{u}$, and the covariance matrix C. Then the minimum variance portfolio is given by,*

$$\mathbf{w}^{mvp} = \frac{\mathbf{u} C^{-1}}{\mathbf{u} C^{-1} \mathbf{u}^\top}.$$

Finally, the expected return for the minimum variance portfolio is given by,

$$\mu_P^{mvp} = \mu \mathbf{w}^{mvp\top}.$$

4.3.3   The Efficient Frontier for Portfolio of n Risky Assets

We now elaborate on the concept of "efficient frontier". In the preceding discussion, we determined the minimum variance portfolio, without imposing any specification on the expected returns. However, it is possible that the minimum variance portfolio results in a corresponding expected return that falls below the target return of the investor. In order to accommodate this concern of the investor, the notion of efficient frontier has to be introduced.

Definition 4.3.4 *(Efficient Portfolio)* An efficient portfolio is a feasible portfolio which either has more return than every other portfolio with the same risk, or has less risk than any other portfolio with the same return. The collection of efficient portfolios is called the efficient frontier.

This means that the efficient portfolios are formed by portfolios satisfying either of the following criteria:

(1) Among all the portfolios with identical σ_P, the portfolio with maximum μ_P.
(2) Among all the portfolios with identical μ_P, the portfolio with minimum σ_P.

It will be observed graphically later on that it suffices to determine the weight of portfolio satisfying one of the criterion 1 or 2. Accordingly, we will consider the problem of minimization of σ_P, or equivalently minimization of σ_P^2, subject to the given level of return $\mu_P = \mu \mathbf{w}^\top$, for a portfolio P of n assets. We define the Lagrangian as,

$$G(\mathbf{w}, \lambda_1, \lambda_2) = \mathbf{w}C\mathbf{w}^\top + \lambda_1\left(1 - \mathbf{u}\mathbf{w}^\top\right) + \lambda_2\left(\mu_P - \mu\mathbf{w}^\top\right).$$

Taking the partial derivative of $G(\mathbf{w}, \lambda_1, \lambda_2)$ with respect to each w_i ($i = 1, 2, \ldots, n$), λ_1 and λ_2, we obtain,

$$C\mathbf{w}^\top = \frac{\lambda_1}{2}\mathbf{u}^\top + \frac{\lambda_2}{2}\mu^\top, \ \mathbf{u}\mathbf{w}^\top = 1 \text{ and } \mu\mathbf{w}^\top = \mu_P.$$

Now,

$$\mathbf{w}C = \mathbf{w}C^\top = \frac{\lambda_1}{2}\mathbf{u} + \frac{\lambda_2}{2}\mu \Rightarrow \mathbf{w} = \left(\frac{\lambda_1}{2}\mathbf{u} + \frac{\lambda_2}{2}\mu\right)C^{-1}.$$

Substituting this $\mathbf{w}$ in $\mathbf{u}\mathbf{w}^\top = 1$ and $\mu\mathbf{w}^\top = \mu_P$, we obtain,

$$\mathbf{u}\left[\left(\frac{\lambda_1}{2}\mathbf{u} + \frac{\lambda_2}{2}\mu\right)C^{-1}\right]^\top = 1 \Rightarrow (\mathbf{u}C^{-1}\mathbf{u}^\top)\lambda_1 + \left(\mathbf{u}C^{-1}\mu^\top\right)\lambda_2 = 2,$$

and

$$\mu\left[\left(\frac{\lambda_1}{2}\mathbf{u} + \frac{\lambda_2}{2}\mu\right)C^{-1}\right]^\top = \mu_P \Rightarrow (\mu C^{-1}\mathbf{u}^\top)\lambda_1 + \left(\mu C^{-1}\mu^\top\right)\lambda_2 = 2\mu_P.$$

We now define, $A_1 = \mathbf{u}C^{-1}\mu^\top$, $A_2 = \mu C^{-1}\mu^\top$, $A_3 = \mathbf{u}C^{-1}\mathbf{u}^\top$, and $A_4 = A_2 A_3 - A_1^2$. Then the system of equations becomes

$$A_3\lambda_1 + A_1\lambda_2 = 2 \text{ and } A_1^\top\lambda_1 + A_2\lambda_2 = 2\mu_P.$$

Solving this system of linear equations in two unknowns λ_1 and λ_2, we obtain,

$$\frac{\lambda_1}{2} = \frac{\begin{vmatrix} A_2 & \mu_P \\ A_1 & 1 \end{vmatrix}}{\begin{vmatrix} A_2 & A_1 \\ A_1 & A_3 \end{vmatrix}} = \frac{A_2 - A_1\mu_P}{A_4},$$

and

$$\frac{\lambda_2}{2} = \frac{\begin{vmatrix} \mu_P & A_1 \\ 1 & A_3 \end{vmatrix}}{\begin{vmatrix} A_2 & A_1 \\ A_1 & A_3 \end{vmatrix}} = \frac{A_3 \mu_P - A_1}{A_4}.$$

Hence we get,

$$\mathbf{w} = \left(\frac{\lambda_1}{2} \mathbf{u} + \frac{\lambda_2}{2} \mu \right) C^{-1} = \left(\frac{A_2 - A_1 \mu_P}{A_4} \right) \mathbf{u} C^{-1} + \left(\frac{A_3 \mu_P - A_1}{A_4} \right) \mu C^{-1}.$$

Defining $\mathbf{f_1} = \dfrac{A_2}{A_4} \mathbf{u} C^{-1} - \dfrac{A_1}{A_4} \mu C^{-1}$ and $\mathbf{f_2} = -\dfrac{A_1}{A_4} \mathbf{u} C^{-1} + \dfrac{A_3}{A_4} \mu C^{-1}$, we get

$$\mathbf{w} = \mathbf{f_1} + \mathbf{f_2} \mu_P.$$

Theorem 4.3.5 (The Efficient Frontier) *Consider a portfolio of n assets, with unit vector* $\mathbf{u}$*, return vector* μ*, and covariance matrix* C*. Then the efficient frontier (or efficient portfolios) is given by,*

$$\mathbf{w}^{ef} = \mathbf{f_1} + \mathbf{f_2} \mu_P,$$

where $\mathbf{f_1}$ *and* $\mathbf{f_2}$ *have already been defined in terms of* $\mathbf{u}$*,* μ*, and* C*.*

The theorem essentially says that a portfolio $\mathbf{w}^{ef}$ on the efficient frontier is a linear function of μ_P. Since $\mathbf{f_1}$ and $\mathbf{f_2}$ are same for all portfolio on the efficient frontier, therefore the efficient portfolio for a given level of return μ_P follows immediately from this linear relation. This means that as an investor, once the n assets have been decided upon and the expected level of return μ_P has been specified, then the efficient portfolio for the investor results from the linear formula for the weights. Further, it can be shown that any portfolio on the efficient frontier is equivalent to investments in two efficient portfolios with unequal expected return. Let P_1 and P_2 be two portfolios on the efficient frontier with the respective expected returns being μ_{P_1} and μ_{P_2} $(\mu_{P_1} \neq \mu_{P_2})$. Let us construct a portfolio, P of P_1 and P_2, with the respective weights being, α_1 and α_2, where $\alpha_1 + \alpha_2 = 1$. Then,

$$\mathbf{w}_P = \alpha_1 \mathbf{w}_{P_1} + \alpha_2 \mathbf{w}_{P2} \text{ and } r_P = \alpha_1 r_{P_1} + \alpha_2 r_{P_2}.$$

Now we get,

$$\mu_P = E(r_P) = \alpha_1 \mu_{P_1} + \alpha_2 \mu_{P_2}.$$

Also, since P_1 and P_2 are efficient portfolios, therefore,

$$\mathbf{w}_{P_1} = \mathbf{f_1} + \mathbf{f_2} \mu_{P_1} \text{ and } \mathbf{w}_{P_2} = \mathbf{f_1} + \mathbf{f_2} \mu_{P_2}.$$

Hence,

$$\begin{aligned} \mathbf{w}_P &= \alpha_1 \left(\mathbf{f_1} + \mathbf{f_2} \mu_{P_1} \right) + \alpha_2 \left(\mathbf{f_1} + \mathbf{f_2} \mu_{P_2} \right) \\ &= \mathbf{f_1}(\alpha_1 + \alpha_2) + \mathbf{f_2} \left(\alpha_1 \mu_{P_1} + \alpha_2 \mu_{P_2} \right) \\ &= \mathbf{f_1} + \mathbf{f_2} \mu_P. \end{aligned}$$

This leads us to the following theorem.

Theorem 4.3.6 (The Two Fund Separation Theorem) *A portfolio on the efficient frontier can be constructed using a linear combination of two portfolios on the efficient frontier.*

Let $\mathbf{w}_P$ be a portfolio on the efficient frontier. Therefore, we can write $\mathbf{w}_P = \mathbf{f_1} + \mathbf{f_2}\mu_P$, which results in the variance of this portfolio, on the efficient frontier, being given by,

$$\sigma_P^2 = \mathbf{w}_P C \mathbf{w}_P^\top = (\mathbf{f_1} + \mathbf{f_2}\mu_P) C (\mathbf{f_1} + \mathbf{f_2}\mu_P)^\top = \frac{A_3}{A_4} \left(\mu_P - \frac{A_1}{A_3} \right)^2 + \frac{1}{A_3}.$$

Thus we have obtained a relation between the variance, σ_P^2 and the expected return, μ_P of the portfolio P on the efficient frontier. In order to explore this geometrically, we consider two cases:

Case 1: We consider the (σ_P^2, μ_P) plane. Then the relation is of the form,

$$x = \frac{A_3}{A_4} \left(y - \frac{A_1}{A_3} \right)^2 + \frac{1}{A_3} \text{ where } x = \sigma_P^2 \text{ and } y = \mu_P.$$

This is a parabola with the vertex $\left(\frac{1}{A_3}, \frac{A_1}{A_3} \right)$, as can be seen in Fig. 4.1.

Case 2: We now consider the important case of the (σ_P, μ_P) plane. Then the relation is of the form,

$$x^2 = \frac{A_3}{A_4} \left(y - \frac{A_1}{A_3} \right)^2 + \frac{1}{A_3} \text{ where } x = \sigma_P \text{ and } y = \mu_P.$$

This is a hyperbola with the vertex $\left(\frac{1}{\sqrt{A_3}}, \frac{A_1}{A_3} \right)$, which has the asymptotes

of $\mu_P = \frac{A_1}{A_3} \pm \sqrt{\frac{A_4}{A_3}} \sigma_P$, as can be seen in Fig. 4.2.

Note that the minimum variance portfolio is given by the point $\left(\frac{1}{\sqrt{A_3}}, \frac{A_1}{A_3} \right)$, in the (σ_P, μ_P) plane.

Finally, we consider a portfolio P on the efficient frontier, with P^* being a managed portfolio. Then, the covariance between them is given by,

$$\sigma_{PP^*} = w_{P^*} C w_P^\top = (f_1 + f_2\mu_{P^*}) C (f_1 + f_2\mu_P)^\top$$

$$= \frac{A_3}{A_4} \left(\mu_P - \frac{A_1}{A_3} \right) \left(\mu_{P^*} - \frac{A_1}{A_3} \right) + \frac{1}{A_3}.$$

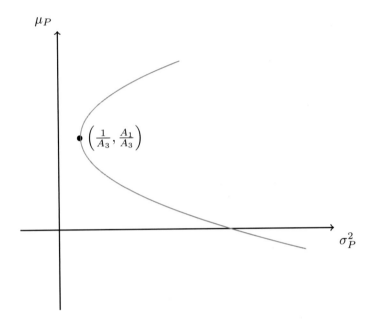

Fig. 4.1 (σ_P^2, μ_P) diagram for Case 1

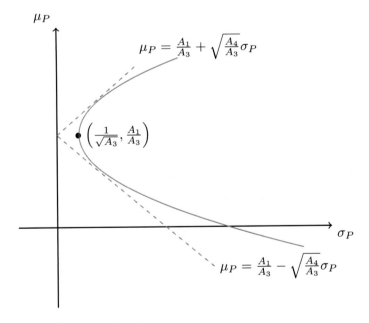

Fig. 4.2 (σ_P, μ_P) diagram for Case 2

4.3.4 The Efficient Frontier for Portfolio of n Risky Assets and a Riskfree Asset

We now consider a portfolio P of n risky assets, along with a riskfree asset. As before, the return and weight of the j-th risky asset are denoted by r_j and w_j, respectively. Let the return and weight of the newly introduced riskfree asset be denoted by μ_f and w_f, respectively. Note that the w_j's considered here are different from the w_i's considered for a portfolio of exclusively risky assets, since the weight w_f has to be now accommodated in the constraint. The expected return and variance of the portfolio P are given by,

$$\mu_P = \boldsymbol{\mu}_{P*}\mathbf{w_P}^{\top} + w_f\mu_f,$$

and

$$\sigma_P^2 = \mathbf{w_{P*}}C\mathbf{w_{P*}}^{\top} + w_f^2\sigma_f^2 + 2\mathbf{w_{P*}}^{\top}\mathbf{w}_f\sigma_{P*f} = \mathbf{w_{P*}}C\mathbf{w}_{P*}^{\top},$$

where P^* indicates the collection of all the risky assets (but excluding the riskfree asset) of the portfolio P^*. Also we have the statutory constraint that $w_f = 1 - \mathbf{w}_{P*}^{\top}\mathbf{u}$. Consequently, the determination of the efficient frontier, in this scenario, involves the minimization of σ_P or equivalently the minimization of σ_P^2, subject to a given level of return,

$$\mu_P = \boldsymbol{\mu}_{P*}\mathbf{w}_{P*}^{\top} + \mu_f(1 - \mathbf{w}_{P*}^{\top}\mathbf{u}),$$

for the portfolio P^* of the n risky assets, along with that of a riskfree asset. We define the Lagrangian as,

$$G(\mathbf{w}_{P*}, \lambda) = \mathbf{w}_{P*}^{\top}C\mathbf{w}_{P*} + \lambda\left[\mu_P - \boldsymbol{\mu}_P\mathbf{w}_{P*}^{\top} - \mu_f(1 - \mathbf{w}_{P*}^{\top}\mathbf{u})\right].$$

Note that here we have taken only one Lagrangian multiplier, since there is only one constraint. Taking the partial derivative of $G(\mathbf{w}_{P*}, \lambda)$ with respect to each w_j ($j = 1, 2, \ldots, n$), and λ, we obtain,

$$2C\mathbf{w}_{P*} - \lambda\left(\boldsymbol{\mu}_{P*} - \mu_f\mathbf{u}\right) = 0 \text{ and } \mu_P = \boldsymbol{\mu}_{P*}\mathbf{w}_{P*}^{\top} + \mu_f(1 - \mathbf{w}_{P*}^{\top}\mathbf{u}).$$

Now,

$$2C\mathbf{w}_{P*} = \lambda\left(\boldsymbol{\mu}_{P*} - \mu_f\mathbf{u}\right) \Rightarrow \mathbf{w}_{P*} = \frac{\lambda}{2}C^{-1}\left(\boldsymbol{\mu}_{P*} - \mu_f\mathbf{u}\right).$$

Substituting $\mathbf{w}_{P*}$ in $\mu_P = \boldsymbol{\mu}_{P*}\mathbf{w}_{P*}^{\top} + \mu_f(1 - \mathbf{w}_{P*}^{\top}\mathbf{u})$, we obtain

$$\mu_P = \mu_f + \mathbf{w}_{P*}^{\top}\left(\boldsymbol{\mu}_{P*} - \mu_f\mathbf{u}\right) = \mu_f + \frac{\lambda}{2}\left(\boldsymbol{\mu}_{P*} - \mu_f\mathbf{u}\right)^{\top}C^{-1}\left(\boldsymbol{\mu}_{P*} - \mu_f\mathbf{u}\right).$$

We define,

$$A = \left(\boldsymbol{\mu}_{P*} - \mu_f\mathbf{u}\right)^{\top}C^{-1}\left(\boldsymbol{\mu}_{P*} - \mu_f\mathbf{u}\right) = A_2 - 2\mu_f A_1 + \mu_f^2 A_3.$$

Hence,

$$\mu_P = \mu_f + \frac{\lambda}{2}A \Rightarrow \frac{\lambda}{2} = \frac{\mu_P - \mu_f}{A}.$$

Therefore, we obtain,

$$\mathbf{w}_{P*} = \frac{\lambda}{2}C^{-1}\left(\boldsymbol{\mu}_{P*} - \mu_f\mathbf{u}\right) = \left(\frac{\mu_P - \mu_f}{A}\right)C^{-1}\left(\boldsymbol{\mu}_{P*} - \mu_f\mathbf{u}\right).$$

Consequently, $w_f = (1 - \mathbf{w}_{P*}^{\top}\mathbf{u})$, and, we get,

$$\mathbf{w}_{P*} = -C^{-1}\left(\boldsymbol{\mu}_{P*} - \mu_f\mathbf{u}\right)\left(\frac{\mu_f}{A}\right) + C^{-1}\left(\boldsymbol{\mu}_{P*} - \mu_f\mathbf{u}\right)\left(\frac{1}{A}\right)\mu_P.$$

Defining

$$\mathbf{g}_1 = -C^{-1}\left(\boldsymbol{\mu}_{P*} - \mu_f\mathbf{u}\right)\left(\frac{\mu_f}{A}\right) \text{ and } \mathbf{g}_2 = C^{-1}\left(\boldsymbol{\mu}_{P*} - \mu_f\mathbf{u}\right)\left(\frac{1}{A}\right),$$

we get,

$$\mathbf{w} = \mathbf{g}_1 + \mathbf{g}_2\mu_P.$$

Thus we have the theorem:

Theorem 4.3.7 (The Efficient Frontier) *Consider a portfolio P of n risky assets and one riskfree asset, with the unit vector $\mathbf{u}$, return vector $\boldsymbol{\mu}$ of risky assets, covariance matrix C of risky assets. Then the efficient frontier is given by,*

$$\mathbf{w} = \mathbf{g}_1 + \mathbf{g}_2\mu_P,$$

where $\mathbf{g}_1$ and $\mathbf{g}_2$ have already been defined in terms of $\mathbf{u}$, $\boldsymbol{\mu}_{P*}$, and C.

Let $\mathbf{w}_P$ be a portfolio P on the efficient frontier. Then we can write,

$$\mathbf{w}_{P*} = \left(\frac{\mu_P - \mu_f}{A}\right)C^{-1}\left(\boldsymbol{\mu}_{P*} - \mu_f\mathbf{u}\right),$$

which results in the variance of this portfolio on the efficient frontier being given by

$$\sigma_P^2 = \mathbf{w}_{P*}^{\top}C\mathbf{w}_{P*} = \left(\boldsymbol{\mu}_{P*} - \mu_f\mathbf{u}\right)^{\top}C^{-1}\left(\boldsymbol{\mu}_{P*} - \mu_f\mathbf{u}\right)\left(\frac{\mu_P - \mu_f}{A}\right)^2$$

$$= A\left(\frac{\mu_P - \mu_f}{A}\right)^2 = \frac{\left(\mu_P - \mu_f\right)^2}{A}.$$

This gives us (since σ_P must be non-negative),

$$\sigma_P = \begin{cases} \frac{(\mu_P - \mu_f)}{\sqrt{A}} & \text{when } \mu_P \geq \mu_f \\ -\frac{(\mu_P - \mu_f)}{\sqrt{A}} & \text{when } \mu_P < \mu_f, \end{cases}$$

that is,

$$\mu_P = \mu_f \pm \sqrt{A}\sigma_P.$$

Thus the efficient frontier is a pair of straight line in the (σ_P, μ_P), plane with the slope(s) $\pm\sqrt{A}$ and intercept μ_f, and can be seen in Fig. 4.3.

Example 4.3.8 Consider a portfolio of two assets a_1 and a_2 with the expected returns of $\mu_1 = 4\%$ and $\mu_2 = 6\%$, respectively. Further $\sigma_1 = 6\%$ and $\sigma_2 = 8\%$, and the assets are uncorrelated. If w_1^{min} and w_2^{min} are the weights of a_1 and a_2 respectively, at which the portfolio attains the minimum variance, then determine $w_1^{min} - w_2^{min}$. What is the expected value of this portfolio.

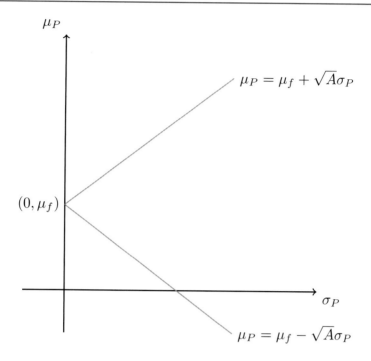

Fig. 4.3 Efficient frontier with a riskfree asset

Here,

$$w_2^{\min} = \frac{\sigma_1^2 - \sigma_{12}}{\sigma_1^2 + \sigma_2^2 - 2\sigma_{12}} = \frac{\sigma_1^2}{\sigma_1^2 + \sigma_2^2} = \frac{0.06^2}{(0.06)^2 + (0.08)^2} = 0.36.$$

Therefore, $w_1^{\min} = 0.64$. Hence, $w_1^{\min} - w_2^{\min} = 0.64 - 0.36 = 0.28$. Thus the expected value of the minimum variance portfolio is, $\mu_1 w_1^{\min} + \mu_2 w_2^{\min} = 0.04 \times 0.64 + 0.06 \times 0.36 = 0.0472$.

Example 4.3.9 Consider two efficient portfolios P_1 and P_2 with $\mu_{P_1} = 0.04$ and $\mu_{P_2} = 0.09$. If α and $(1 - \alpha)$ are the weights of investment in P_1 and P_2, respectively, then determine the value of $\dfrac{\alpha}{1 - \alpha}$, which achieves an overall expected return of 6%.

Let P denote the combined portfolios. Then by the Two-Fund theorem

$$\mu_P = 0.04\alpha + 0.09(1 - \alpha) = 0.06 \Rightarrow \alpha = 0.6 \Rightarrow \frac{\alpha}{1 - \alpha} = \frac{0.6}{0.4} = 1.5.$$

4.4   Capital Asset Pricing Model

The capital market theory is the backbone of what is known as the Capital Asset Pricing Model (CAPM–pronounced "cap"-"M". We begin with the following basic underlying assumptions:

(1) Investors are assumed to make their investment decisions based only on the mean-variance analysis of the terminal wealth, or the rate of returns.
(2) Investors are assumed to be risk-averse, who are driven by the goal of maximization of utility (to be defined later) of the terminal wealth, or the rate of returns.
(3) All assets (including human capital) are assumed to be marketable, as well as divisible (which means that fractional units of assets can be bought or sold).
(4) It is assumed that there are no transaction costs and no tax considerations.
(5) The assumption is that there exists a riskfree rate, and all lending and borrowing take place at the riskfree rate.
(6) There exists a perfect capital market scenario, wherein all information is assumed to be transparently available, there is no margin requirements, and there is no restrictions on transaction, including lending and borrowing, as well as short selling.
(7) All investors are assumed to be operating over identical time horizons, with homogeneous expectation, and identical opinion and perception, vis-a-vis the estimation of expected return, variance, and covariance.

In this section, we deal with two important lines, namely the capital market line (CML) and the security market line (SML) or CAPM.

4.4.1   Capital Market Line

Recall that for a portfolio comprising of only risky assets, the variance (or equivalently, the standard deviation) is always positive. In contrast, the variance (or equivalently, the standard deviation) for a riskfree asset is zero. We now consider a holding (or a long position) in a risky portfolio (or asset) and a riskfree asset. Let the weights of risky portfolio (asset) and the riskfree asset be w_1 and $w_2 = 1 - w_1$, respectively. Then, the expected return of the resulting portfolio is given by,

$$\mu_P = w_1 \mu_1 + (1 - w_1)\mu_f.$$

Here, $\mu_1 = E(r_1)$ is the expected return of the risky portfolio (or asset), and μ_f is the riskfree rate of return. The variance of the resulting portfolio is given by,

$$\sigma_P^2 = w_1^2 \sigma_1^2 + (1 - w_1)^2 \sigma_f^2 + 2w_1 (1 - w_1) \sigma_{1f},$$

where σ_1^2 is the variance of the risky portfolio (or asset), σ_f^2 is the variance of riskfree asset and σ_{1f} is the covariance of return of the risky portfolio (or asset) and the riskfree rate. Since the riskfree rate is constant, therefore $\sigma_f^2 = 0$ and $\sigma_{1f} = 0$.

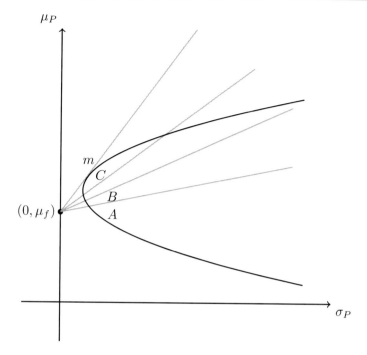

Fig. 4.4 Capital market line (CML)

Consequently, $\sigma_P^2 = w_1^2 \sigma_1^2$. Taking the positive square root we obtain, $\sigma_P = w_1 \sigma_1$, which implies that $w_1 = \dfrac{\sigma_P}{\sigma_1}$. Substituting this value of w_1, we get,

$$\mu_P = \frac{\sigma_P}{\sigma_1}\mu_1 + \left(1 - \frac{\sigma_P}{\sigma_1}\right)\mu_f \Rightarrow \mu_P = \mu_f + \left(\frac{\mu_1 - \mu_f}{\sigma_1}\right)\sigma_P.$$

This is a line in the (σ_P, μ_P) plane, passing through the points $(0, \mu_f)$ and (σ_1, μ_1). In other words, this equation constitutes a family of lines emanating from $(0, \mu_f)$, as shown in Fig. 4.4.

From the Figure, it can be observed that it is most desirable to invest in the portfolio, where the slope of the family of lines is maximum, that is, the line joining $(0, \mu_f)$ and the portfolio m (which is called the tangency portfolio or the market portfolio), in which case the equation of this line becomes,

$$\mu_P = \mu_f + \left(\frac{\mu_m - \mu_f}{\sigma_m}\right)\sigma_P,$$

which is called the CML.

The weight of the market portfolio can be derived as follows: We maximize the slope of all the lines in the family, subject to the sum of weights being equal to 1, i.e.,

$$\text{Maximize } \frac{\mu_P - \mu_f}{\sigma_P} = \frac{\mathbf{w}\boldsymbol{\mu}^\top - \mu_f}{\sqrt{\mathbf{w}C\mathbf{w}^\top}} \text{ such that } \mathbf{wu}^\top = 1.$$

Accordingly, we define the Lagrangian,

$$H(\mathbf{w}, \lambda) = \frac{\mathbf{w}\boldsymbol{\mu}^\top - \mu_f}{\sqrt{\mathbf{w}C\mathbf{w}^\top}} + \lambda \left(1 - \mathbf{w}\mathbf{u}^\top\right).$$

Differentiating with respect to each w_i ($i = 1, 2, \ldots, n$) and λ, we obtain,

$$\frac{\mu_P - \mu_f}{\sigma_P^2}\mathbf{w}C = \boldsymbol{\mu} - \lambda\sigma_P\mathbf{u} \Rightarrow \lambda = \frac{\mu_f}{\sigma_P}.$$

Therefore we have,

$$\frac{\mu_P - \mu_f}{\sigma_P^2}\mathbf{w}C = \boldsymbol{\mu} - \mu_f\mathbf{u}$$

$$\Rightarrow \left(\frac{\mu_P - \mu_f}{\sigma_P^2}\right)\mathbf{w} = \left(\boldsymbol{\mu} - \mu_f\mathbf{u}\right)C^{-1}$$

$$\Rightarrow \frac{\mu_P - \mu_f}{\sigma_P^2} = \left(\frac{\mu_P - \mu_f}{\sigma_P^2}\right)\mathbf{w}\mathbf{u}^\top = \left(\boldsymbol{\mu} - \mu_f\mathbf{u}\right)C^{-1}\mathbf{u}^\top$$

$$\Rightarrow \mathbf{w} = \frac{\left(\boldsymbol{\mu} - \mu_f\mathbf{u}\right)C^{-1}}{\left(\boldsymbol{\mu} - \mu_f\mathbf{u}\right)C^{-1}\mathbf{u}^\top}.$$

Thus we have the following Theorem.

Theorem 4.4.1 (Weight of Market Portfolio) *The weights of the market portfolio are given by,*

$$\mathbf{w} = \frac{\left(\boldsymbol{\mu} - \mu_f\mathbf{u}\right)C^{-1}}{\left(\boldsymbol{\mu} - \mu_f\mathbf{u}\right)C^{-1}\mathbf{u}^\top}.$$

4.4.2   Security Market Line or CAPM

Consider a portfolio comprising of an investment in a risky asset, with weight $w_{1'}$ and the remaining weight $w_m = 1 - w_{1'}$ being invested in the market portfolio. The resulting portfolio, P' comprises of the market portfolio assets, along with excess demand for one risky asset. The expected return and standard deviation of the portfolio P' are,

$$\mu_{P'} = w_{1'}\mu_{1'} + (1 - w_{1'})\mu_m$$

$$\sigma_{P'} = \left(w_{1'}^2\sigma_{1'}^2 + \left(1 - w_{1'}\right)^2\sigma_m^2 + 2w_{1'}\left(1 - w_{1'}\right)\sigma_{1'm}\right)^{\frac{1}{2}},$$

respectively. The slope in the $(\sigma_{P'}, \mu_{P'})$ plane is given by,

$$\frac{d\mu_{P'}}{d\sigma_{P'}} = \frac{\frac{d\mu_{P'}}{dw_{1'}}}{\frac{d\sigma_{P'}}{dw_{1'}}} = \frac{\mu_{1'} - \mu_m}{w_{1'}\sigma_{1'}^2 - \left(1 - w_{1'}\right)\sigma_m^2 + \left(1 - 2w_{1'}\right)\sigma_{1'm}} \times \sigma_{P'}.$$

The slope for the portfolio P' when it comprises exclusively of the market portfolio is given by,

$$\left.\frac{d\mu_{P'}}{d\sigma_{P'}}\right|_{w_{1'}=0} = \frac{\left(\mu_{1'} - \mu_m\right)\sigma_{P'}}{\sigma_{1'm} - \sigma_m^2}.$$

But this being the slope of the tangent line at m must be equal to the slope of the CML, given by $\dfrac{\mu_m - \mu_f}{\sigma_m}$. Therefore,

$$\frac{\mu_m - \mu_f}{\sigma_m} = \left(\frac{\mu_{1'} - \mu_m}{\sigma_{1'm} - \sigma_m^2}\right)\sigma_{P'} \Rightarrow \mu_{1'} = \mu_f + \left(\frac{\mu_m - \mu_f}{\sigma_m^2}\right)\sigma_{1'm}.$$

Recall that $\beta_{1'} = \dfrac{\sigma_{1'm}}{\sigma_m^2}$. Therefore,

$$\mu_{1'} = \mu_f + \left(\mu_m - \mu_f\right)\beta_{1'}.$$

In general, we get $\mu = \mu_f + (\mu_m - \mu_f)\beta$ which is the SML or CAPM.

4.4.3 Pricing Aspects

We consider a risky asset during a single period. Suppose that the asset is purchased at time $t = 0$, for price p_0, and its value at time $t = 1$ is a random variable p_1, whose expected value is μ_{p_1}. Then, expected return on this asset is given by,

$$\mu = E\left[\frac{p_1 - p_0}{p_0}\right] = \frac{\mu_{p_1}}{p_0} - 1.$$

A low μ is equivalent to a high p_0, and vice versa. In equilibrium $\mu = \mu_f + \left(\mu_m - \mu_f\right)\beta$. Now, we consider two scenarios:

(1) If p_0 is low, then μ is higher than what is given by SML or CAPM, i.e., $\mu > \mu_f + \left(\mu_m - \mu_f\right)\beta$.
(2) If p_0 is high, then μ is lower than what is given by SML or CAPM, i.e., $\mu < \mu_f + \left(\mu_m - \mu_f\right)\beta$.

Thus, we can surmise that, if $\mu > \mu_f + \left(\mu_m - \mu_f\right)\beta$, then the asset is under priced (p_0 is low), and if $\mu < \mu_f + \left(\mu_m - \mu_f\right)\beta$, then the asset is overpriced (p_0 is high). Consequently, the price will go up or come down, respectively, until the CAPM equilibrium of $\mu = \mu_f + \left(\mu_m - \mu_f\right)\beta$ is attained. In equilibrium,

$$\frac{\mu_{p_1}}{p_0} - 1 = \mu_f + \left(\mu_m - \mu_f\right)\beta \Rightarrow p_0 = \frac{\mu_{p_1}}{1 + \mu_f + \left(\mu_m - \mu_f\right)\beta},$$

which is a pricing equation for the initial price p_0.

4.4.4 Single Index Model

A very simple model for return of an asset is the return generating function, or the single index model, given by,

$$r_i = \alpha_i + \beta_i r_m + \epsilon_i.$$

Here, r_i is the return from the i-th asset and r_m is the return from a market index. Further, α_i and β_i are constants, to be determined from the historical data and ϵ_i is a random variable, known as the residual error. The single index model makes the following assumptions:

(1) The expected value of the residual error is zero, i.e., $E(\epsilon_i) = 0$.
(2) The residual error and the market return are independent of each other, i.e., $\mathrm{Cov}(\epsilon_i, r_m) = 0$.
(3) The residual errors of two different assets i and j are independent, i.e., $\mathrm{Cov}(\epsilon_i, \epsilon_j) = 0$.

Let the variance of r_m and ϵ_i be denoted by σ_m^2 and $\sigma_{\epsilon_i}^2$, respectively. Now,

$$E(r_i) = E(\alpha_i + \beta_i r_m + \epsilon_i) = \alpha_i + \beta_i E(r_m), \text{ since } E(\epsilon_i) = 0.$$

and

$$\begin{aligned}
\mathrm{Var}(r_i) &= \mathrm{Var}(\alpha_i + \beta_i r_m + \epsilon_i) \\
&= \mathrm{Var}(\alpha_i) + \mathrm{Var}(\beta_i r_m) + \mathrm{Var}(\epsilon_i) + \mathrm{Cov}(\alpha_i, \beta_i r_m) \\
&\quad + \mathrm{Cov}(\alpha_i, \epsilon_i) + \mathrm{Cov}(\beta_i r_m, \epsilon_i) \\
&= 0 + \beta_i^2 \mathrm{Var}(r_m) + \mathrm{Var}(\epsilon_i) + 0 + 0 + 0 \\
&= \beta_i^2 \sigma_m^2 + \sigma^2 \epsilon_i.
\end{aligned}$$

Also,

$$\begin{aligned}
\mathrm{Cov}(r_i, r_j) &= E\left[(r_i - E(r_i))(r_j - E(r_j))\right] \\
&= E\left[(\beta_i(r_m - E(r_m)) + \epsilon_i)(\beta_j(r_m - E(r_m)) + \epsilon_j)\right] \\
&= E\left[\beta_i \beta_j(r_m - E(r_m))^2\right] + E\left[\beta_i(r_m - E(r_m))\epsilon_j\right] \\
&\quad + E\left[\beta_j(r_m - E(r_m))\epsilon_i\right] + E(\epsilon_i \epsilon_j) \\
&= \beta_i \beta_j \sigma_m^2 + 0 + 0 + 0 = \beta_i \beta_j \sigma_m^2.
\end{aligned}$$

Hence we have the theorem:

Theorem 4.4.2 *For the single index model* $r_i = \alpha_i + \beta_i r_m + \epsilon_i$, *we have,*

(i) $E(r_i) = \alpha_i + \beta_i E(r_m)$.
(ii) $\mathrm{Var}(r_i) = \beta_i^2 \sigma_m^2 + \sigma^2 \epsilon_i$.
(iii) $\mathrm{Cov}(r_i, r_j) = \beta_i \beta_j \sigma_m^2 \text{ for } i \neq j$.

If the random variable r_i takes the values r_{it}, $t = 1, 2, \ldots, T$ and r_m takes the values r_{mt}, $t = 1, 2, \ldots, T$, then we have the estimates,

$$E(r_i) = \frac{1}{T} \sum_{t=1}^{T} r_{it} \text{ and } E(r_m) = \frac{1}{T} \sum_{t=1}^{T} r_{mt}.$$

Then the parameter estimates based on historical data are

$$\widehat{\beta_i} = \frac{\sum_{t=1}^{T} (r_{it} - E(r_i))(r_{mt} - E(r_m))}{\sum_{t=1}^{T} (r_{it} - E(r_i))^2},$$

$$\widehat{\alpha_i} = E(r_i) - \widehat{\beta_i} E(r_m),$$

$$\widehat{\sigma}_{\epsilon_i}^2 = \frac{1}{T-1} \sum_{t=1}^{T} \left(r_{it} - (\widehat{\alpha_i} + \widehat{\beta_i} r_m)\right)^2.$$

A variation of the single index model results in the following single index model for excess return $\tilde{r}_i = r_i - \mu_f$.

$$\tilde{r}_i = \tilde{\alpha}_i + \beta_i \tilde{r}_m + \tilde{\epsilon}_i,$$

where $\tilde{r}_m = r_m - \mu_f$ and $\tilde{\alpha}_i = \alpha_i - \mu_f(1 - \beta_i)$. Accordingly,

$$E(\tilde{r}_i) = \tilde{\alpha}_i + \beta_i E(\tilde{r}_m) \text{ and } \text{Var}(\tilde{r}_i) = \beta_i^2 \sigma_m^2 + \sigma_{\epsilon_i}^2.$$

We now consider the single index model in the paradigm of a portfolio P comprising of n assets. Let r_i be the return of the i-th asset which is modeled by $r_i = \alpha_i + \beta_i r_m + \epsilon_i$, with the corresponding weight being w_i. Then the return on the portfolio is,

$$r_p = \sum_{i=1}^{n} w_i r_i = \sum_{i=1}^{n} w_i (\alpha_i + \beta_i r_m + \epsilon_i)$$

$$= \sum_{i=1}^{n} w_i \alpha_i + \left(\sum_{i=1}^{n} w_i \beta_i\right) r_m + \sum_{i=1}^{n} w_i \epsilon_i$$

$$= \alpha_p + \beta_p r_m + \epsilon_p,$$

where $\alpha_p = \sum_{i=1}^{n} w_i \alpha_i$, $\beta_p = \sum_{i=1}^{n} w_i \beta_i$, and $\epsilon_p = \sum_{i=1}^{n} w_i \epsilon_i$. Note that the relation $\beta_p = \sum_{i=1}^{n} w_i \beta_i$ can be proved using the definitions of β_i and β_p. It can also be proved that $E(r_p) = \alpha_p + \beta_P E(r_m)$ and $\text{Var}(r_p) = \beta_p^2 \sigma_m^2 + \sigma^2 \epsilon_p$. Hence the risk of the portfolio given by $\text{Var}(r_p)$ is decomposed into the systematic risk or non-diversifiable risk, $\beta_p^2 \sigma_m^2$ and unsystematic or diversifiable risk $\sigma_{\epsilon_p}^2$. This means that one can potentially get rid of the diversifiable risk, by diversifying the portfolio.

4.4.5 Multi-index Models

We now extend the single index model to a multi-index setup with K indices to have the model for return of the asset as,

$$r_i = \alpha_i + \sum_{k=1}^{K} \beta_{ik} I_k + \epsilon_i.$$

Here, α_i and β_{ik} for $k = 1, 2, \ldots, K$ are constants, to be determined. As before, ϵ_i is the residual error. The multi-index model makes the following assumptions:

(1) The expected value of the residual error is zero, i.e., $E(\epsilon_i) = 0$.
(2) The residual error and the indices are independent of each other, i.e., Cov $(\epsilon_i, I_k) = 0$ for $k = 1, 2, \ldots, K$.
(3) The indices are independent of each other, i.e., $\text{Cov}(I_k, I_l) = 0 \; \forall \; k, l = 1, 2, \ldots, K, \; k \neq l$.
(4) The residual error of two different assets i and j are independent, i.e., Cov $(\epsilon_i, \epsilon_j) = 0$.

Let the variance of I_k and ϵ_i be denoted by σ_k^2 and $\sigma_{\epsilon_i}^2$, respectively. In practice, the indices may not be independent, but can be made independent of each other using a methodology of orthogonalization. Accordingly, without loss of generality, we make assumption that, $\text{Cov}(I_k, I_l) = 0 \; \forall \; k, l = 1, 2, \ldots, K, \; k \neq l$. Now, we get,

$$E(r_i) = E\left(\alpha_i + \sum_{k=1}^{K} \beta_{ik} I_k + \epsilon_i\right) = \alpha_i + \sum_{k=1}^{K} \beta_{ik} E(I_k),$$

and

$$\text{Var}(r_i) = \text{Var}\left(\alpha_i + \sum_{k=1}^{K} \beta_{ik} I_k + \epsilon_i\right)$$

$$= \text{Var}(\alpha_i) + \text{Var}\left(\sum_{k=1}^{K} \beta_{ik} I_k\right) + \text{Var}(\epsilon_i) + \text{Cov}\left(\alpha_i, \sum_{k=1}^{K} \beta_{ik} I_k\right)$$

$$+ \text{Cov}(\alpha_i, \epsilon_i) + \text{Cov}\left(\sum_{k=1}^{K} \beta_{ik} I_k, \epsilon_i\right)$$

$$= 0 + \sum_{k=1}^{K} \beta_{ik}^2 \sigma_k^2 + \sigma_{\epsilon_i}^2 + 0 + 0 + 0 = \sum_{k=1}^{K} \beta_{ik}^2 \sigma_k^2 + \sigma_{\epsilon_i}^2.$$

Also,

$$\text{Cov}(r_i, r_j) = E\left[(r_i - E(r_i))(r_j - E(r_j))\right]$$

$$= E\left[\left(\sum_{k=1}^{K} \beta_{ik}(I_k - E(I_k)) + \epsilon_i\right)\left(\sum_{k=1}^{K} \beta_{jk}(I_k - E(I_k)) + \epsilon_j\right)\right]$$

$$= E\left[\sum_{k=1}^{K} \beta_{ik}\beta_{jk}(I_k - E(I_k))^2\right] + E\left[\sum_{k=1}^{K} \beta_{ik}(I_k - E(I_k))\epsilon_j\right]$$

$$+ E\left[\sum_{k=1}^{K} \beta_{jk}(I_k - E(I_k))\epsilon_i\right] + E(\epsilon_i\epsilon_j)$$

$$= \sum_{k=1}^{K} \beta_{ik}\beta_{jk}\sigma_k^2 + 0 + 0 + 0 = \sum_{k=1}^{K} \beta_{ik}\beta_{jk}\sigma_k^2.$$

Hence we have the theorem,

Theorem 4.4.3 *For the multi-index model* $r_i = \alpha_i + \sum_{k=1}^{K} \beta_{ik}I_k + \epsilon_i,$

(i) $E(r_i) = \alpha_i + \sum_{k=1}^{K} \beta_{ik}E(I_k).$
(ii) $Var(r_i) = \sum_{k=1}^{K} \beta_{ik}^2\sigma_k^2 + \sigma_{\epsilon_i}^2.$
(iii) $Cov(r_i, r_j) = \sum_{k=1}^{K} \beta_{ik}\beta_{jk}\sigma_k^2.$

Example 4.4.4 In the CAPM framework, suppose that market return and risk (as given by standard deviation of returns) are 7.75% and 12%, respectively, with the riskfree rate being 4%. If you are willing to accept a risk of 15% on a managed portfolio, then determine the expected return (in percentage) of this portfolio.

Here, $\mu_m = 0.0775$, $\sigma_m = 0.12$, $\mu_f = 0.04$ and $\sigma_P = 0.15$. Then

$$\mu_P = 0.04 + \frac{0.0775 - 0.04}{0.12} \times 0.015 = 0.08688 \text{ or } 8.688\%.$$

Example 4.4.5 Consider an asset a_i with $\sigma_i = 10\%$, $\sigma_{im} = 0.576\%$, and 8%. Determine the diversifiable risk, as given by σ_{ϵ_i} (in percentage).

Here $\beta_i = \dfrac{\sigma_{im}}{\sigma_m^2} = \dfrac{0.00576}{(0.08)^2} = 0.9$. Therefore,

$$\sigma_{\epsilon_i}^2 = \sigma_i^2 - \beta_i^2\sigma_m^2 = 0.1^2 - 0.9^2 \times 0.08^2 = 0.004816 \Rightarrow \sigma_{\epsilon_i} = 0.0694 \text{ or } 6.94\%.$$

Example 4.4.6 Consider an asset a_i with $\mu_f = 5\%$, $\mu_m = 8\%$ and $\beta_i = 0.5$. Determine the value of μ_i (in percentage), for the asset to be correctly priced.

In order to be correctly priced, the asset has to satisfy SML. Therefore,

$$\mu_i = 0.05 + (0.08 - 0.05) \times 0.5 = 0.065 \text{ or } 6.5\%.$$

4.5 Arbitrage Pricing Theory

An alternative to CAPM is offered by what is known as the Arbitrage Pricing Theory (APT). The APT is more realistic than CAPM, from the perspective that it does not make the assumption of all investors making their decisions based on the mean-variance framework, but rather that the investors are driven by the preference for greater returns, over less returns. We begin our discussion on APT with a simple case, where the returns on the i-th asset are modeled using a single index model (without the residual error term) as,

$$r_i = \alpha_i + \beta_i I_1,$$

where I_1 is the index. The notion of APT here is to preclude the possibility of riskless profit or arbitrage. We now consider two assets, namely i (already introduced) and j, for which the model is $r_j = \alpha_j + \beta_j I_1$. Further, we require that $\beta_i \neq \beta_j$ (else the response of the asset returns to the index would be identical). Suppose that we construct a portfolio P of the i-th and j-th asset, with the respective weights being w_i and $w_j = 1 - w_i$. Then the return of the portfolio P is given by,

$$r_p = [w_i \alpha_i + (1 - w_i)\alpha_j] + [w_i \beta_i + (1 - w_i)\beta_j]I_1.$$

The source of risk in this model is the index I_1, and hence in order to eliminate this, we choose the coefficient of I_1 to be zero i.e.,

$$w_i \beta_i + (1 - w_i)\beta_j = 0 \Rightarrow w_i = \frac{\beta_j}{\beta_j - \beta_i} \text{ (Recall that } \beta_i \neq \beta_j \text{)}.$$

Then the return of the portfolio becomes,

$$r_p = w_i \alpha_i + (1 - w_i)\alpha_j = \frac{\alpha_i \beta_j - \alpha_j \beta_i}{\beta_j - \beta_i}.$$

Now this return r_p becomes a return, free from any risk (or randomness), and hence by the no-arbitrage principle, it must equal the riskfree rate $\mu_f := \lambda_0$ (say). Then,

$$\frac{\alpha_i \beta_j - \alpha_j \beta_i}{\beta_j - \beta_i} = \lambda_0 \Rightarrow (\beta_j - \beta_i)\lambda_0 = \alpha_i \beta_j - \alpha_j \beta_i \Rightarrow \frac{\alpha_j - \lambda_0}{\beta_j} = \frac{\alpha_i - \lambda_0}{\beta_i} = k_1,$$

where k_1 is some constant. This shows that,

$$\alpha_i = \lambda_0 + \beta_i k_1 \text{ and } \alpha_j = \lambda_0 + \beta_j k_1.$$

Hence in general,

$$r_i = \lambda_0 + \beta_i k_1 + \beta_i I_1 \Rightarrow E(r_i) = \lambda_0 + \beta_i (k_1 + E(I_1)) \Rightarrow E(r_i) = \lambda_0 + \beta_i \lambda_1,$$

where $\lambda_1 := k_1 + E(I_1)$. Recall that the CAPM is given by,

$$E(r_i) - \mu_f = \beta_i (E(r_m) - \mu_f).$$

Accordingly, the CAPM is a particular case of the APT, with $\lambda_0 = \mu_f$ and $\lambda_1 = E(r_m) - \mu_f$. We now extend the APT to a general setup.

Theorem 4.5.1 *Consider n risky assets, with the return of the i-th asset being governed by the relation,*

$$r_i = \alpha_i + \sum_{k=1}^{K} \beta_{ik} I_k,$$

for $i = 1, 2, \ldots, n$ *and* $k = 1, 2, \ldots, K$, *with* $K < n$. *Here* I_k *is the k-th factor. Then there exists constant* $\lambda_0, \lambda_1, \ldots, \lambda_k$, *such that,*

$$E(r_i) = \lambda_0 + \sum_{k=1}^{K} \beta_{ik} \lambda_k \ for \ i = 1, 2, \ldots, n.$$

This theorem can be proved under the following conditions:

$$\sum_{i=1}^{n} w_i \beta_{ik} = 0 \text{ and } \sum_{i=1}^{n} w_i E(r_i) = \mu_f.$$

Example 4.5.2 Consider two assets a_i and a_j with

$$r_i = \alpha_i + 2I_1 + I_2,$$
$$r_j = \alpha_j + 7I_1 + 4I_2.$$

Determine the values of λ_1 and λ_2, with $E(r_i) = 10\%$, $E(r_j) = 12\%$, and $\mu_f = 10\%$.

Here, $\beta_{i1} = 2$, $\beta_{i2} = 1$, $\beta_{j1} = 7$, and $\beta_{j2} = 4$. Therefore,

$$0.10 = 0.1 + 2\lambda_1 + \lambda_2,$$
$$0.12 = 0.1 + 7\lambda_1 + 4\lambda_2.$$

Solving we get $\lambda_1 = -0.02$ and $\lambda_2 = 0.04$.

4.6   Variations of CAPM

We now consider two variations of the CAPM.

4.6.1   Black's Zero-Beta Model

The Black's zero-beta model is a result of reconsideration of the assumptions about the existence of a riskfree asset with a single riskfree rate, at which all the borrowing and lending takes place. Accordingly, the riskfree asset is replaced by an asset or portfolio, with the beta being zero, in the CAPM framework. Assuming that this zero-beta portfolio lies on the efficient frontier, on which the market portfolio lies, we create a portfolio P with weight s in the zero-beta portfolio, z, and the remaining weight $(1-s)$ in the market portfolio m. Then the expected return and risk of portfolio P are,

$$E(r_P) = sE(r_z) + (1 - s)E(r_m).$$

$$\sigma_P = \left(s^2\sigma_z^2 + (1-s)^2\sigma_m^2 + 2s(1-s)\sigma_{zm}\right)^{\frac{1}{2}} = \left(s^2\sigma_z^2 + (1-s)^2\sigma_m^2\right)^{\frac{1}{2}},$$

respectively. Then,

$$\frac{dE(r_P)}{d\sigma_P}\bigg|_{s=0} = \frac{E(r_m) - E(r_z)}{\sigma_m}.$$

This line intersects μ_P axis at the point $E(r_z)$. Therefore,

$$\frac{E(r_P) - E(r_z)}{\sigma_P} = \frac{E(r_m) - E(r_z)}{\sigma_m},$$

which implies,

$$E(r_P) = E(r_z) + \left(\frac{E(r_m) - E(r_z)}{\sigma_m}\right)\sigma_P.$$

Now, we consider a portfolio made of a risky asset and market portfolio. Then the slope of the point m is $\dfrac{(E(r_i) - E(r_m))\sigma_m}{\sigma_{im} - \sigma_m^2}$ which is equal to $\left(\dfrac{E(r_m) - E(r_z)}{\sigma_m}\right)$.

This gives,

$$\frac{(E(r_i) - E(r_m))\sigma_m}{\sigma_{im} - \sigma_m^2} = \left(\frac{E(r_m) - E(r_z)}{\sigma_m}\right),$$

which implies that,

$$E(r_i) = E(r_z) + \frac{\sigma_{im}}{\sigma_m^2}(E(r_m) - E(r_z)) = E(r_z) + \beta_i(E(r_m) - E(r_z)).$$

4.6.2   Brennan's After-Tax Model

The Brennans's After-Tax Model takes into account the taxes on both the sources of return on a risky asset, namely dividend yield and capital gains. Let r_{ji} denotes the return on the j-th asset being held by investor i. Further, let t_{dy_i} and t_{cg_i} be constants denoting the tax rate on the dividend yield and the capital gain, respectively for investor i. Then, the after-tax return, for investor i, on j-th risky security (with dividend δ_j), and a riskfree asset are,

$$r_{ji}^{(a)} = (r_j - \delta_j)(1 - t_{cg_i}) + \delta_j(1 - t_{dy_i}), \tag{4.1}$$

$$\mu_{fi}^{(a)} = \mu_f(1 - t_{dy_i}). \tag{4.2}$$

This gives

$$r_{ji}^{(a)} - \mu_{fi}^{(a)} = (r_j - \delta_j)(1 - t_{cg_i}) + (\delta_j - \mu_f)(1 - t_{dy_i}). \tag{4.3}$$

Now

$$\text{Var}\left(r_{ji}^{(a)}\right) = (1 - t_{cg_i})^2\sigma_j^2, \tag{4.4}$$

and

$$\text{Cov}\left(r_{ji}^{(a)}, r_{ki}^{(a)}\right) = (1 - t_{cg_i})^2 \sigma_{jk}. \tag{4.5}$$

Let m^* be the market like portfolio in this situation. In a market in equilibrium, we have a CML like relation,

$$E\left(r_{ji}^{(a)}\right) = \mu_{fi}^{(a)} + \left(\frac{E\left(r_{m^*i}^{(a)}\right) - \mu_{fi}^{(a)}}{\text{Var}\left(r_{m^*i}^{(a)}\right)}\right) \text{Cov}\left(r_{m^*i}^{(a)}, r_{ji}^{(a)}\right). \tag{4.6}$$

Substituting (4.1)–(4.5) in (4.6) we get,

$$[E(r_j) - \mu_f]\left[1 - t_{cg_i}\right] = \left(\delta_j - \mu_f\right)\left(t_{dy_i} - t_{cg_i}\right)$$
$$+ \left[\frac{E\left(r_{m^*i}^{(a)}\right) - \mu_{fi}^{(a)}}{\text{Var}\left(r_{m^*i}^{(a)}\right)}\right] \text{Cov}\left(r_{m^*i}^{(a)}, r_{ji}^{(a)}\right).$$

Dividing both sides by $\left(1 - t_{cg_i}\right)$, we obtain

$$[E(r_j) - \mu_f] - (\delta_j - \mu_f)\left(\frac{t_{dy_i} - t_{cg_i}}{1 - t_{cg_i}}\right) = \left(\frac{E\left(r_{m^*i}^{(a)}\right) - \mu_{fi}^{(a)}}{\text{Var}\left(r_{m^*i}^{(a)}\right)\left(1 - t_{cg_i}\right)}\right) \text{Cov}\left(r_{m^*i}^{(a)}, r_{ji}^{(a)}\right).$$

Let the absolute amount invested by investor i be denoted by W_i and let $\theta_i = \dfrac{E\left(r_{m^*i}^{(a)}\right) - \mu_{fi}^{(a)}}{\text{Var}\left(r_{m^*i}^{(a)}\right)\left(1 - t_{cg_i}\right)}$. Then we obtain,

$$\frac{W_i}{\theta_i}\left[E(r_j) - \mu_f\right] - \frac{W_i}{\theta_i}\left(\delta_j - \mu_f\right)\left(\frac{t_{dy_i} - t_{cg_i}}{1 - t_{cg_i}}\right) = W \cdot \text{Cov}\left(r_{m^*i}^{(a)}, r_{ji}^{(a)}\right),$$

which when summed up, for all K investors, i.e, for $i = 1, 2, \ldots, K$, gives,

$$\sum_{i=1}^{K} \frac{W_i}{\theta_i}\left[E(r_j) - \mu_f\right] - \sum_{i=1}^{K} \frac{W_i}{\theta_i}\left(\delta_j - \mu_f\right)\left(\frac{t_{dy_i} - t_{cg_i}}{1 - t_{cg_i}}\right)$$
$$= \sum_{i=1}^{K} W_i \text{Cov}\left(r_{m^*i}^{(a)}, r_{ji}^{(a)}\right)$$
$$= W \sum_{i=1}^{K} \text{Cov}\left(\frac{W_i r_{m^*i}^{(a)}}{W}, r_j\right)$$
$$= W \text{Cov}(r_m, r_j),$$

where W is the value of the market portfolio. Define,

$$B_1 = \frac{W}{\sum_{i=1}^{K} \frac{W_i}{\theta_i}} \quad \text{and} \quad B_2 = \frac{B_1}{W}\left[\sum_{i=1}^{K} \frac{W_i}{\theta_i}\left(\frac{t_{dy_i} - t_{cg_i}}{1 - t_{cg_i}}\right)\right].$$

Then,

$$[E(r_j) - \mu_f] - B_2\left[\delta_j - \mu_f\right] = B_1 \text{Cov}\left(r_m, r_j\right),$$

which implies,

$$B_1 = \frac{\left[E(r_j) - \mu_f\right] - B_2\left(\delta_j - \mu_f\right)}{\mathrm{Cov}\left(r_m, r_j\right)} = \frac{\left[E(r_m) - \mu_f\right] - B_2\left(\delta_m - \mu_f\right)}{\sigma_m^2}, \quad \text{when } j = m.$$

Substituting, we obtain,

$$E(r_j) = \mu_f + B_2\left(\delta_j - \mu_f\right) + \left[\frac{\left(E(r_m) - \mu_f\right) - B_2\left(\delta_m - \mu_f\right)}{\sigma_m^2}\right]\mathrm{Cov}\left(r_m, r_j\right),$$

which implies that,

$$E(r_j) = \mu_f + B_2\left(\delta_j - \mu_f\right) + \left[\left(E(r_m) - \mu_f\right) - B_2\left(\delta_m - \mu_f\right)\right]\beta_j.$$

Note that in general, $0 < t_{cg_i} < t_{dy_i} < 1$. Also, if there are no taxes, i.e., $t_{cg_i} = 0 = t_{dy_i}$, the above relation reduces to the SML or CAPM.

4.7 Portfolio Performance Analysis

We now present three single-parameter portfolio performance measure, due to Sharpe, Treynor, and Jensen, driven by the consideration of both return and risk.

1. **Sharpe's ratio**: Let μ_f and $E(r_p)$ be the riskfree return and expected return of an actively managed portfolio P, respectively. Then, the excess return or risk premium (additional expected return over the riskfree return, for the investor, resulting from getting into the risky investment) is given by $E(r_P) - \mu_f$. Further, let σ_P denote the risk for P. Then the Sharpe's ratio is defined as,

$$S_P = \frac{E(r_P) - \mu_f}{\sigma_P},$$

   which is a single-index performance indicator that can be used to rank the desirability of several actively managed portfolios. Graphically, S_P is the slope of the straight line $E(r_P) = \mu_f + S_P\sigma_P$ in the $(\sigma_P, E(r_P))$ plane.

2. **Treynor's ratio**: In this case, the portfolio beta (β_P) is taken as the measure of systematic risk and (as specified earlier), is given by the weighted sum of the betas of the constituent assets of the portfolio. Then the Treynor's ratio is given by,

$$T_P = \frac{E(r_P) - \mu_f}{\beta_P},$$

   and provides a measure of the desirability of actively managed portfolios.

3. **Jensen's Alpha**: The expected return of a portfolio in the CAPM framework is given by

$$E(r_P) = \mu_f + \left[E(r_m) - \mu_f\right]\beta_P.$$

Jensen sought to estimate the performance of a portfolio in terms of excess return achieved by the portfolio, over the return predicted by the CAPM, through the introduction of Jensen's Alpha,

$$\alpha_P = E(r_P) - \left(\mu_f + \left[E(r_m) - \mu_f\right]\beta_P\right),$$

or equivalently,

$$E(r_P) = \alpha_P + \mu_f + \left[E(r_m) - \mu_f\right]\beta_P.$$

Note that α_P is a measure of disequilibrium.

We now carry out, a comparative perspective of Sharpe's ratio, Treynor's ratio, and Jensen's Alpha. Accordingly, we begin with the equation of Jensen's Alpha,

$$E(r_P) - \mu_f = \alpha_P + \left[E(r_m) - \mu_f\right]\beta_P.$$

Dividing both sides by β_P, we obtain,

$$T_P = \frac{E(r_P) - \mu_f}{\beta_P} = \frac{\alpha_P}{\beta_P} + \left[E(r_m) - \mu_f\right].$$

This shows that the Treynor's ratio (T_P) is a linear transformation of Jensen's Alpha (α_P). Recall that $\beta_P = \frac{\rho_{Pm}\sigma_P\sigma_m}{\sigma_m^2}$. In case of a well-diversified portfolio, we have, $\rho_{Pm} \approx 1$, which gives $\beta_P \approx \frac{\sigma_P}{\sigma_m}$. This results in,

$$E(r_P) - \mu_f \approx \alpha_P + \left[E(r_m) - \mu_f\right]\frac{\sigma_P}{\sigma_m}.$$

Dividing both sides by σ_P, we obtain,

$$S_P = \frac{E(r_P) - \mu_f}{\sigma_P} \approx \frac{\alpha_P}{\sigma_P} + \frac{E(r_m) - \mu_f}{\sigma_m}.$$

Thus, Sharpe's ratio (S_P) is approximately a linear transformation of Jensen's Alpha (α_P). Finally, for a well-diversified portfolio with $\beta_P \approx \frac{\sigma_P}{\sigma_m}$, we have,

$$T_P \approx \frac{E(r_P) - \mu_f}{\beta_P} = \frac{E(r_P) - \mu_f}{\sigma_P}\sigma_m = S_P\sigma_m.$$

The Treynor's ratio is approximately a linear transformation of the Sharpe's ratio. We conclude our discussion on portfolio performance analysis, through a discussion on performance decomposition.

1. **Fama's Decomposition**: This decomposition is driven by two components, namely selectively and risk, motivated by the comparison of the performance of an actively managed portfolio, with that of two benchmark portfolios from the SML. Let the actively managed portfolio be denoted by P, with the corresponding risk and systematic risk being σ_P and β_P, respectively. We now take two portfolios on the SML denoted Q_1 and Q_2, satisfying the condition $\beta_{Q_1} = \beta_P$ and $\beta_{Q_2} = \sigma_P$. Since Q_1 is on the SML, therefore,

$$E(r_{Q_1}) = \mu_f + \beta_{Q_1}\left[E(r_m) - \mu_f\right] \ \left(\beta_{Q_1} = \beta_P\right).$$

Also, since Q_2 is on the SML, therefore,

$$E(r_{Q_2}) = \mu_f + \beta_{Q_2}\left[E(r_m) - \mu_f\right] \ (\beta_{Q_2} = \sigma_P).$$

Now, the performance of the portfolio P is decomposed into,

$$E(r_P) - \mu_f = \left[E(r_P) - E(r_{Q_1})\right] + \left[E(r_{Q_1}) - \mu_f\right].$$

(A) Now, we consider first term (selectively),

$$E(r_P) - E(r_{Q_1}) = \left[E(r_P) - E(r_{Q_2})\right] + \left[E(r_{Q_2}) - E(r_{Q_1})\right].$$

The term

$$E(r_P) - E(r_{Q_1}) = \left[E(r_P) - \mu_f\right] - \sigma_P\left[E(r_m) - \mu_f\right],$$

gives the net selectivity. Also the term

$$E(r_{Q_2}) - E(r_{Q_1}) = (\sigma_P - \beta_P)\left(E(r_m) - \mu_f\right),$$

is the diversification term.

(B) Now, we consider the second term (risk). Let P_0 be the portfolio reflecting the investor level of risk appetite. Then,

$$E\left(r_{Q_1}\right) - \mu_f = \left[E(r_{Q_1}) - E(r_{P_0})\right] + \left[E(r_{P_0}) - \mu_f\right].$$

Here, the terms $E(r_{Q_1}) - E(r_{P_0})$ and $E(r_{P_0}) - \mu_f$ represent the manager's risk and investor's risk, respectively.

2. **Brinson's Model for Performance Decomposition**: This model is based on three stages of active portfolio management, namely investment policy, timing, and asset picking. In order to assess the performance of the portfolio manager, we consider an actively managed portfolio, denoted by P and a benchmark portfolio, denoted by B. We consider K number of asset classes for both the portfolios. Let w_{P_i} and r_{P_i} denote the weight and return, respectively of the i-th class in the actively managed portfolio P. Similarly, let w_{B_i} and r_{B_i} denote the weight and return, respectively of the i-th class in the benchmark portfolio B. Then, the excess return is given by,

$$r_P - r_B = \sum_{i=1}^{K} w_{P_i} r_{P_i} - \sum_{i=1}^{K} w_{B_i} r_{B_i},$$

which can be decomposed into,

$$r_P - r_B = \sum_{i=1}^{K}\left(w_{P_i} - w_{B_i}\right)r_{B_i} + \sum_{i=1}^{K} w_{B_i}\left(r_{P_i} - r_{B_i}\right) + \sum_{i=1}^{K}\left(w_{P_i} - w_{B_i}\right)\left(r_{P_i} - r_{B_i}\right).$$

The first term $\sum_{i=1}^{K}\left(w_{P_i} - w_{B_i}\right)r_{B_i}$ represents the timing. The second term $\sum_{i=1}^{K} w_{B_i}\left(r_{P_i} - r_{B_i}\right)$ represents the asset picking. Finally, the term $\sum_{i=1}^{K}\left(w_{P_i} - w_{B_i}\right)\left(r_{P_i} - r_{B_i}\right)$ is the interaction term between allocation and asset pricing.

Example 4.7.1 Consider two risky assets a_i and a_j with $\mu_i = 6\%$, $\mu_j = 8\%$, $\beta_i = 0.12$, $\beta_j = 0.18$, and $\mu_f = 4\%$. If T_i and T_j are the Treynor ratios of a_i and a_j, respectively, then determine $|T_i - T_j|$.

We have,

$$|T_i - T_j| = \left| \frac{0.06 - 0.04}{0.12} - \frac{0.08 - 0.04}{0.18} \right| = 0.0556.$$

Example 4.7.2 For a portfolio P, $\mu_P = 8\%$, $\sigma_P = 17.5\%$, and $\mu_f = 4\%$. Further $\mu_m = 8\%$ and $\sigma_m = 12\%$. If the Jensen's Alpha $\alpha_P = 1\%$, then determine the Treynor's ratio T_P and the Sharpe's ratio S_P, for P.

For Jensen's Alpha,

$$\mu_P - \mu_f = \alpha_P + \beta_P[\mu_m - \mu_f] \Rightarrow 0.08 - 0.04 = 0.01 + \beta_P[0.08 - 0.04] \Rightarrow \beta_P = 0.75.$$

Therefore,

$$T_P = \frac{\mu_P - \mu_f}{\beta_P} = \frac{0.08 - 0.04}{0.75} = 0.0533,$$

and

$$S_P = \frac{\mu_P - \mu_f}{\sigma_P} = \frac{0.08 - 0.04}{0.175} = 0.2286.$$

Example 4.7.3 Consider the performance of a managed portfolio P and benchmark portfolio B as given in the table below. Determine the contribution of asset allocation and security selection toward the excess return of P over B.

Class (i)	Return (r_{P_i})	Weight (w_{P_i})
Stocks	9.20%	0.75
Bonds	3.10%	0.20
Cash	1.20%	0.05

Class (i)	Return (r_{B_i})	Weight (w_{B_i})
Stocks	7.30%	0.60
Bonds	2.80%	0.25
Cash	1.20%	0.15

The contribution of asset allocation is,

$$\sum_i (w_{P_i} - w_{B_i}) r_{B_i} = 0.00835 \text{ and } 0.835\%.$$

The contribution of securities selection is

$$\sum_i (r_{P_i} - r_{B_i}) w_{B_i} = 0.01215 \text{ and } 1.215\%.$$

4.8 Exercise

Exercise 4.1 Consider an asset a_i, with the return r_i, whose values under three different economic scenarios s_1, s_2, and s_3 are tabulated below. Determine the variance of returns $Var(r_i)$ (in percentage).

Scenario	Return (r_i)	Probability
s_1	8%	0.25
s_2	10%	0.50
s_3	12%	0.25

Solution: Here, $E(r_i) = 0.25 \times 0.08 + 0.50 \times 0.1 + 0.25 \times 0.12 = 0.1$ or 10%. Therefore, $Var(r_i) = 0.25 \times (0.08 - 0.1)^2 + 0.50 \times (0.1 - 0.1)^2 + 0.25 \times (0.12 - 0.1)^2 = 0.002$ or 0.2%.

Exercise 4.2 Consider a portfolio comprising of three assets a_i, $i = 1, 2, 3$, with weights $w_1 = 0.5$, $w_2 = 0.2$, $w_3 = 0.3$, expected returns $\mu_1 = 6.8\%$, $\mu_2 = 5\%$, $\mu_3 = 4\%$ and covariance matrix (in percentage)

$$\begin{bmatrix} 10.5 & 8.3 & 7.2 \\ 8.3 & 9.6 & 6.6 \\ 7.2 & 6.6 & 8.4 \end{bmatrix}.$$

Determine the expected return and variance of this portfolio.

Solution: The expected return is, $0.5 \times 0.068 + 0.2 \times 0.05 + 0.3 \times 0.04 = 0.056$ or 5.6%. Hence, the variance of returns is,

$$0.5^2 \times 0.105 + 0.2^2 \times 0.096 + 0.3^2 \times 0.084$$
$$+ 2 \times 0.5 \times 0.2 \times 0.083 + 2 \times 0.5 \times 0.3 \times 0.072 + 2 \times 0.2 \times 0.3 \times 0.066$$
$$= 0.0838 \text{ or } 8.38\%.$$

Exercise 4.3 The six historical prices of a stock are given by 100, 90, 85, 95, 105, and 110. Estimate the expected return and variance of the stock return based on these historical prices.

Solution: The returns are $r_1 = -0.1$, $r_2 = -0.0556$, $r_3 = 0.1176$, $r_4 = 0.1053$, $r_5 = 0.0476$. Therefore the expected return is $\frac{1}{5}(-0.1 - 0.0556 + 0.1176 + 0.1053 + 0.0476) = 0.02298$. Hence the variance of returns is

$$\frac{1}{5}\left((-0.1 - 0.02298)^2 + (-0.0556 - 0.02298)^2 + (0.1176 - 0.02298)^2\right.$$
$$\left. + (0.1053 - 0.02298)^2 + (0.0476 - 0.02298)^2\right) = 0.010893.$$

Exercise 4.4 Consider a portfolio of two assets a_1 and a_2 with the expected return $\mu_1 = 5\%$ and $\mu_2 = 4\%$, respectively. Further $\sigma_1 = 6\%$ and $\sigma_2 = 5\%$ and the assets are uncorrelated. If $w_1^{\min}$ and $w_2^{\min}$ are the weights of a_1 and a_2 respectively, at which the portfolio attains the minimum variance, then determine $\dfrac{w_1^{\min}}{w_2^{\min}}$. Hence determine the expected value of this portfolio.

Solution: Here

$$w_2^{\min} = \frac{\sigma_1^2 - \sigma_{12}}{\sigma_1^2 + \sigma_2^2 - 2\sigma_{12}} = \frac{\sigma_1^2}{\sigma_1^2 + \sigma_2^2} = \frac{0.06^2}{(0.06)^2 + (0.05)^2} = 0.59.$$

Therefore, $w_1^{\min} = 0.41$. Hence, $\dfrac{w_1^{\min}}{w_2^{\min}} = \dfrac{0.59}{0.41} = 1.439$. Finally, the expected value of the minimum variance portfolio is

$$w_1^{\min}\mu_1 + w_2^{\min}\mu_2 = 0.41 \times 0.05 + 0.59 \times 0.04 = 0.0441 \text{ or } 4.41\%.$$

Exercise 4.5 Consider two efficient portfolios, P_1 and P_2, with the respective weights of

$$\mathbf{w_1} = \begin{bmatrix} 0.3 \\ 0.7 \end{bmatrix} \text{ and } \mathbf{w_2} = \begin{bmatrix} 0.8 \\ 0.2 \end{bmatrix}.$$

A new efficient portfolio P is created with weights $\alpha = 0.4$ and $(1 - \alpha) = 0.6$ being assigned to P_1 and P_2, respectively. If P has the weight $\mathbf{u} = \begin{bmatrix} u_1 \\ u_2 \end{bmatrix}$, then determine $u_1 - u_2$.

Solution: By the Two-Fund Theorem

$$\mathbf{u} = \alpha\mathbf{w_1} + (1 - \alpha)\mathbf{w_2} = 0.4 \begin{bmatrix} 0.3 \\ 0.7 \end{bmatrix} + 0.6 \begin{bmatrix} 0.8 \\ 0.2 \end{bmatrix} = \begin{bmatrix} 0.6 \\ 0.4 \end{bmatrix}.$$

Therefore $u_1 - u_2 = 0.2$.

Exercise 4.6 If $\mu_m = 2\mu_f$ and $\sigma_P = 1.5\sigma_m$, with $\mu_f = 5\%$, then determine μ_P (in percentage).

Solution: From the CML, we get,

$$\mu_P = \mu_f + (\mu_m - \mu_f)\frac{\sigma_P}{\sigma_m} = \mu_f + (2\mu_f - \mu_f)\frac{1.5\sigma_m}{\sigma_m} = 2.5\mu_f = 2.5 \times 5\% = 12.5\%.$$

Exercise 4.7 Consider the following data given in the table.
    If the single index model is used with $r_i = \alpha_i + \beta_i r_m + \epsilon_i$, then determine α_i and β_i using the least square approximation.

Asset Return (r_i) %	Market Return (r_m) %
12	8
13.2	10
6.8	9
3.8	7
10	4
9.2	6
8	5
9.4	6

Solution: Here,

$$E(r_i) = \frac{1}{T} \sum_{t=1}^{T} r_{it}$$

$$= \frac{1}{8} (0.12 + 0.132 + 0.068 + 0.038 + 0.1 + 0.092 + 0.08 + 0.094) = 0.0905,$$

and

$$E(r_m) = \frac{1}{T} \sum_{t=1}^{T} r_{mt}$$

$$= \frac{1}{8} (0.08 + 0.1 + 0.09 + 0.07 + 0.04 + 0.06 + 0.05 + 0.06) = 0.06875.$$

Then we have the following:

r_i	r_m	$r_i - E(r_i)$	$r_m - E(r_m)$	$(r_i - E(r_i))(r_m - E(r_m))$
0.12	0.08	0.0295	0.01125	0.000331875
0.132	0.10	0.0415	0.03125	0.001296875
0.068	0.09	−0.0225	0.02125	−0.000478125
0.038	0.07	−0.0525	0.00125	−0.000065625
0.10	0.04	0.0095	−0.02875	−0.000273125
0.092	0.06	0.0015	−0.00875	−0.000013125
0.08	0.05	−0.0105	−0.01875	0.000196875
0.094	0.06	0.0035	−0.00875	−0.000030625

$$\widehat{\beta}_i = \frac{\frac{1}{T} \sum_{t=1}^{T} (r_{it} - E(r_i)) (r_{mt} - E(r_m))}{\frac{1}{T} \sum_{t=1}^{T} (r_{it} - E(r_i))^2} = 0.158978583,$$

and

$$\widehat{\alpha}_i = E(r_i) - \widehat{\beta}_i E(r_m) = 0.0905 - 0.158978583 \times 0.06875 = 0.079570222.$$

Exercise 4.8 Consider a single period ($t = 0$ to $t = 1$) asset pricing model whose value at time $t = 1$, can be either 110 or 90, each with probability of $\frac{1}{2}$. Further let $\mu_f = 5\%$, $\beta = 0.2$, and $\mu_m = 9\%$. Then using CAPM, determine the correct value of the asset at time $t = 0$.

Solution: Here $\mu_{P_1} = \frac{1}{2} \times 110 + \frac{1}{2} \times 90 = 100$, $\mu_f = 0.05$, $\mu_m = 0.09$, and $\beta = 0.2$. Therefore, the correct value of the asset at time $t = 0$ is

$$p_0 = \frac{\mu_{P_1}}{1 + \mu_f + (\mu_m - \mu_f)\beta} = \frac{100}{1 + 0.05 + 0.04 \times 0.2} = \frac{100}{1.058} = 94.518.$$

Exercise 4.9 Consider a mutual fund F that invests 50% in an asset a_i (with $\mu_i = 10\%$ and $\sigma_i = 12\%$), and the remaining 50% in a riskfree asset (with $\mu_f = 5\%$). If an investor borrows the riskfree asset and invests in F in order to get return of 15%, then determine the standard deviation of the investment by the investor.

Solution: Let P be the investment in F and the riskfree asset. Now $\mu_F = \frac{1}{2} \times 0.10 + \frac{1}{2} \times 0.05 = 0.075$ or 7.5%, and $\sigma_F = \frac{1}{2} \times 0.12 = 0.06$ or 6%. Therefore,

$$\mu_P = w_f \mu_f + (1 - w_f)\mu_F = 0.15 \Rightarrow w_f \times 0.05 + (1 - w_f) \times 0.075 = 0.15 \Rightarrow w_f = -3 \text{ and } (1 - w_f) = 4.$$

This means that we have weight -3 of riskfree asset, i.e., we have borrowed the riskfree asset. Finally, $\sigma_P = 4 \times 0.06 = 0.24$ or 24%.

Exercise 4.10 Consider two assets a_i and a_j, with $\mu_i = 6\%$, $u_j = 10\%$, $\sigma_j = 18\%$. Also let $\mu_m = 11\%$ and $\sigma_m = 20\%$. Let P be a portfolio comprising of a_i and a_j, with equal weights. If $\beta_P = \frac{1}{2}$, then determine σ_{ϵ_j} (in percentage).

Solution: Here $\beta_P = \frac{1}{2} \times \beta_i + \frac{1}{2} \times \beta_j = \frac{1}{2} \Rightarrow \beta_i + \beta_j = 1$. From the SML for a_i and a_j, we get

$$0.06 - \mu_f = (1 - \beta_j)(0.11 - \mu_f) \text{ and } 0.10 - \mu_f = \beta_j(0.11 - \mu_f),$$

which gives $\mu_f = 0.05$. Therefore, $\beta_j = \frac{5}{6}$. Hence,

$$\sigma_j^2 = \beta_j^2 \sigma_m^2 + \sigma_{\epsilon_j}^2 \Rightarrow \sigma_{\epsilon_j} = \sqrt{\sigma_j^2 - \beta_j^2 \sigma_m^2} = \sqrt{0.18^2 - \left(\frac{5}{6}\right)^2 \times 0.2^2} = 0.06799 \text{ or } 6.799\%.$$

Exercise 4.11 Consider two assets a_i and a_j, with,

$$r_i = \alpha_i + I_1 + 2I_2,$$
$$r_j = \alpha_j + 3I_1 + I_2.$$

Determine the values of λ_1 and λ_2, with $E(r_i) = 18\%$, $E(r_j) = 12\%$, and $\mu_f = 12\%$.

Solution: Here $\beta_{i1} = 1$, $\beta_{i2} = 2$, $\beta_{j1} = 3$, and $\beta_{j2} = 1$. Therefore,

$$0.18 = 0.12 + \lambda_1 + 2\lambda_2,$$
$$0.12 = 0.12 + 3\lambda_1 + \lambda_2.$$

Solving we get $\lambda_1 = -0.012$ and $\lambda_2 = 0.036$.

Exercise 4.12 Consider two assets a_i and a_j, with,

$$r_i = \alpha_i + 2I_1 + I_2,$$
$$r_j = \alpha_j + 3I_1 + 4I_2.$$

Derive the formula for $E(r_k)$ by determining λ_0 and λ_1, with $E(r_i) = 15\%$, $E(r_j) = 20\%$, and $\mu_f = 10\%$.

Solution: Here $\beta_{i1} = 2$, $\beta_{i2} = 1$, $\beta_{k1} = 3$, and $\beta_{k2} = 4$. Therefore

$$0.15 = 0.1 + 2\lambda_1 + \lambda_2,$$
$$0.20 = 0.1 + 3\lambda_1 + 4\lambda_2.$$

Solving we get, $\lambda_1 = 0.02$ and $\lambda_2 = 0.01$. Therefore, $E(r_k) = 0.1 + 0.02\beta_{k1} + 0.01\beta_{k2}$.

Exercise 4.13 For a portfolio P, $\mu_P = 5\%$ and $\sigma_P = 17\%$, with $\mu_f = 3\%$. Further, $\mu_m = 8\%$ and $\sigma_m = 11\%$. If the Jensen's Alpha $\alpha_P = 1.5\%$, then determine the Treynor's ratio T_P and Sharpe's ratio S_P, for P.

Solution: For the Jensen's Alpha,

$$\mu_P - \mu_f = \alpha_P + \beta_P[\mu_m - \mu_f] \Rightarrow 0.05 - 0.03 = 0.015 + \beta_P[0.08 - 0.03] \Rightarrow \beta_P = 0.1.$$

Therefore,

$$T_P = \frac{\mu_P - \mu_f}{\beta_P} = \frac{0.05 - 0.03}{0.1} = 0.2,$$

and

$$S_P = \frac{\mu_P - \mu_f}{\sigma_P} = \frac{0.05 - 0.03}{0.17} = 0.1176.$$

Utility Theory

5

The history of utility functions can be traced back to Bernoulli, who sought to resolve the St. Petersburg Paradox, which can be surmised as the "winner is the one who ends with the most at death" which (as we will see later) translates to maximization of the expected utility. Another perspective to this metaphor could arguably be the position that "the winner is the one who spends the most by death", which (as we will see later) translates to the utility of spending or consumption.

5.1 Basics of Utility Functions

Definition 5.1.1 (*Utility Function*) A utility function is defined as a function or mapping, from a set of consumption choices (or alternatives), or wealth level to real numbers, called utils.

Before we delve into the utility theory, in the paradigm of portfolio optimization, we lay down the basic axioms of utility theory.

(1) Given two alternatives I_1 and I_2, an investor prefers I_1 to I_2 if $U(I_1) > U(I_2)$, prefers I_2 to I_1 if $U(I_2) > U(I_1)$, or is indifferent between I_1 and I_2 if $U(I_1) = U(I_2)$.

(2) The choices are transitive, that is, if an investor prefers I_1 to I_2 ($U(I_1) > U(I_2)$) and prefers I_2 to I_3 ($U(I_2) > U(I_3)$), then by transitivity (resulting from ($U(I_1) > U(I_3)$)), they prefer I_1 to I_3.

(3) If an investor is indifferent between I_1 and I_3 ($U(I_1) = U(I_3)$), then if they prefer I_1 to I_2 ($U(I_1) > U(I_2)$), then they prefer I_3 to I_2 ($U(I_3) > U(I_2)$).

(4) If $U(I_1) > U(I_2) > U(I_3)$, then $\exists$ an investment strategy such that,

$$p_1 U(I_1) + p_3 U(I_3) = U(I_2),$$

where p_1 and p_3 are the respective probabilities of I_1 and I_3, occurring.

© The Author(s), under exclusive license to Springer Nature Singapore Pte Ltd. 2023   69
S. P. Chakrabarty and A. Kanaujiya, *Mathematical Portfolio Theory and Analysis*,
Compact Textbooks in Mathematics, https://doi.org/10.1007/978-981-19-8544-7_5

(5) In an investment alternative I_3 doesn't affect I_1 and I_2, then,

$$U(I_1) > U(I_2) \Rightarrow U(I_1) + U(I_3) > U(I_2) + U(I_3).$$

(6) Finally, in the context of the St. Petersburg Paradox, investors are driven by the goal of maximization of expected utility, given by,

$$E(U) = \sum_{i=1}^{K} p_i U(I_i),$$

where p_i is the probability of the ith investment alternative.

Example 5.1.2 Consider two investment opportunities A and B, for an investor with utility function for wealth $U(w) = w^2$.
Opportunity A: An investment of 100 pays 110 with probability $\frac{2}{3}$ or pays 90 with probability $\frac{1}{3}$.
Opportunity B: An investment of 100 pays 105, for certain.
Then determine the value of the difference $E(U_B(w)) - E(U_A(w))$.

Here, the expected utilities are $E(U_A(w)) = \frac{2}{3} \times (110)^2 + \frac{1}{2} \times (90)^2 =$ 10766.6667 and
$E(U_B(w)) = (105)^2 = 11025$. Hence $E(U_B(w)) - E(U_A(w)) = 258.3333$.

Example 5.1.3 Consider the following table of return distribution of three assets a_1, a_2, and a_3. If the utility function of returns is $U(r) = r^2$, and we denote the expected utility of return of asset a_i by $E(U_i(r))$ for $i = 1, 2, 3$, then determine the value of $\dfrac{E(U_1(r)) + E(U_2(r))}{E(U_3(r))}$.

Return	Probability (a_1)	Probability (a_2)	Probability (a_3)
−5%	0.8	0.3	0
0%	0	0.2	0
5%	0.2	0.5	1.0

The values of the expected utilities are,

$$E(U_1(r)) = 0.8 \times (-0.05)^2 + 0 \times 0^2 + 0.2 \times (0.05)^2 = 0.0025,$$
$$E(U_2(r)) = 0.3 \times (-0.05)^2 + 0.2 \times 0^2 + 0.5 \times (0.05)^2 = 0.002,$$
$$E(U_3(r)) = 0 \times (-0.05)^2 + 0 \times 0^2 + 1 \times (0.05)^2 = 0.0025.$$

Hence the required value is given by, $\dfrac{0.0025 + 0.002}{0.0025} = 1.8.$

5.2   Risk Attitude of Investors

While in the context of financial engineering, the assumption is that an investor is risk averse, we will however in this section delineate the three mutually exclusive, but exhaustive categories of risk appetite of investors, namely risk-averse, risk-seeking, and risk-neutral. In order to illustrate this, we consider two investment alternatives.

(1) **Alternative I_1**: A riskless investment for a single period, with a return of r_1 (deterministic). Then the expected utility for return on I_1 is given by,

$$E(U(r_1)) = U(r_1).$$

(2) **Alternative I_2**: A risky investment for a single period with a return of r_2 (random) is given by,

$$r_2 = \begin{cases} r_1 - \delta \text{ with probability } \frac{1}{2}, \\ r_1 + \delta \text{ with probability } \frac{1}{2}. \end{cases}$$

Then the expected utility for return on I_2 is given by,

$$E(U(r_2)) = \frac{1}{2}U(r_1 - \delta) + \frac{1}{2}U(r_1 + \delta).$$

Note that $E(r_1) = r_1 = E(r_2)$.

We now elaborate on the three kinds of risk appetites of investors:

1. **Risk-averse Investor**: A risk-averse investor is one whose utility function is strictly increasing ($U'(w) > 0$) and strictly concave ($U''(w) < 0$). The question this begets is what would the position of the investor be? Accordingly,

$$
\begin{aligned}
E(U(r_1)) = U(r_1) &= U(E(r_2)) \\
&= U\left(\frac{1}{2}(r_1 - \delta) + \frac{1}{2}(r_1 + \delta)\right) \\
&> \frac{1}{2}U(r_1 - \delta) + \frac{1}{2}U(r_1 + \delta) \text{ (since } U \text{ is strictly concave)} \\
&= E(U(r_2)).
\end{aligned}
$$

   This shows that the investor has the expected utility of I_1 (riskfree) more than the expected utility of I_2 (risky), which means that the investor is risk-averse.

2. **Risk-seeking Investor**: A risk-seeking investor is one whose utility function is strictly increasing ($U'(w) > 0$) and strictly convex ($U''(w) > 0$). Accordingly,

$$
\begin{aligned}
E(U(r_1)) = U(r_1) &= U(E(r_2)) \\
&= U\left(\frac{1}{2}(r_1 - \delta) + \frac{1}{2}(r_1 + \delta)\right) \\
&< \frac{1}{2}U(r_1 - \delta) + \frac{1}{2}U(r_1 + \delta) \text{ (since } U \text{ is strictly convex)} \\
&= E(U(r_2)).
\end{aligned}
$$

   This shows that the investor has the expected utility of I_2 (risky) more than the expected utility of I_1 (riskfree), which means that the investor is risk-seeking.

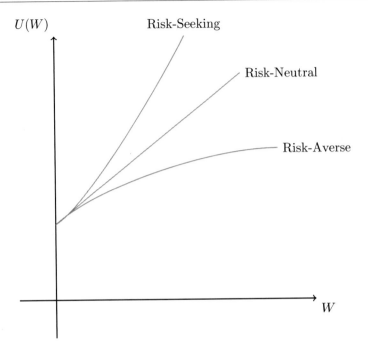

Fig. 5.1 Attitude of investors

3. **Risk-neutral Investor**: A risk-neutral investor is one whose utility function is strictly increasing ($U'(w) > 0$) and is linear ($U''(w) = 0$). Accordingly,

$$E(U(r_1)) = U(r_1) = U(E(r_2))$$
$$= U\left(\frac{1}{2}(r_1 + \delta) + \frac{1}{2}(r_1 - \delta)\right)$$
$$= \frac{1}{2}U(r_1 - \delta) + \frac{1}{2}U(r_1 + \delta) \text{ (since } U \text{ is linear)}$$
$$= E(U(r_2)).$$

This shows that the investor has the expected utility of I_1 (riskfree) equal to the expected utility of I_2 (risky), which means that the investor is risk-neutral.

The three different forms of risk attitude of investors is illustrated in Fig. 5.1.

We can also look at the risk appetite of investor in the mean-variance framework. Accordingly, we take the Taylor series expansion of $U(w)$ about $E(w)$,

$$U(w) = U(E(w) + (w - E(w)))$$
$$= U(E(w)) + U'(E(w))(w - E(w)) + \frac{1}{2!}U''(E(w))(w - E(w))^2$$
$$+ \sum_{k=3}^{\infty} \frac{1}{k!}U^{(k)}(E(w))(w - E(w))^k.$$

In the Markowitz setup, the assumption is that investor's assess an investment on the basis of the first two moments, and all higher-order derivatives, $U^{(k)}(E(w)) = 0$, for $k = 3, 4, \ldots$. This results in (by taking expectation on both sides),

$$E[U(w)] = U(E(w)) + \frac{1}{2}U''(E(w))Var(w).$$

Thus we have that,

$$\frac{1}{2}U''(E(w))Var(w) = E[U(w)] - U[E(w)].$$

Thus for a risk-averse investor,

$$E[U(w)] - U[E(w)] < 0 \iff U''(w) < 0.$$

Similarly, for a risk-seeking investor,

$$E[U(w)] - U[E(w)] > 0 \iff U''(w) > 0.$$

And finally, for a risk-neutral investor,

$$E[U(w)] - U[E(w)] = 0 \iff U''(w) = 0.$$

These three inferences are consistent with the setup of the preceding enumerated description of risk attitude of investors.

Example 5.2.1 Consider the following table of return distribution of three assets a_1, a_2, and a_3. Consider three investors A, B, and C with the respective utility function of returns, $U_A(r) = 2r - r^2$, $U_B(r) = 2r$, and $U_C(r) = 2r + r^2$. Calculate the expected utility for the investor A, B, and C for each of the three assets and determine the asset with the highest expected utility for each of the investors. Hence identify the nature of risk attitude for each of the three investors.

Return	Probability (a_1)	Probability (a_2)	Probability (a_3)
−5%	0.6	0.2	0
0%	0	0	0
5%	0.2	0.2	0
8%	0.2	0.6	1.0

The values of the expected utilities are:

$$E(U_A(r_1)) = -0.0113, \quad E(U_A(r_2)) = 0.0912, \quad E(U_A(r_3)) = 0.1536,$$
$$E(U_B(r_1)) = -0.0080, \quad E(U_B(r_2)) = 0.0960, \quad E(U_B(r_3)) = 0.1600,$$
$$E(U_C(r_1)) = -0.0047, \quad E(U_C(r_2)) = 0.1008, \quad E(U_C(r_3)) = 0.1664.$$

From the above table, we can see that highest expected utility for each investor is in case of asset a_3. Also, we can say that investor A is risk-averse, B is risk-neutral, and investor C is risk-seeking.

5.3   More on Utility Theory

In this section, we introduce the notions of "marginal utility", "certainty equivalent", "absolute-risk-aversion", and "relative-risk-aversion". Marginal utility is defined as the additional utility, an investor derives from a small change of their wealth level. Since it is reasonable to expect that an investor will desire more wealth than less wealth, so the marginal utility of every investor is always positive, irrespective of their risk attitude. Mathematically, this is given by $U^{'}(w) > 0$. In case of risk-averse (risk-seeking) investor, $U^{''}(w) < 0$ ($U^{''}(w) > 0$), which means that the marginal utility is diminishing (increasing).

Given a certain wealth level, w, the "certainty equivalent" is a constant amount of cash, $CE(w)$, which makes the investor indifferent, between a risky investment and $CE = CE(w)$. Mathematically,

$$E\left[U(w)\right] = U(CE).$$

Accordingly, the risk premium is defined as,

$$\text{Risk Premium} := E(w) - CE(w).$$

We take the Taylor series expansion up to second-order terms, to obtain,

$$U(w) = U(E(w) + w - E(w))$$
$$\approx U(E(w)) + (w - E(w))U^{'}(E(w)) + \frac{1}{2!}(w - E(w))^2 U^{''}(E(w)).$$

Taking expectation on both sides, we obtain,

$$E\left[U(w)\right] = U(E(w)) + \frac{1}{2}Var(w)U^{''}(E(w)).$$

Also,

$$E\left[U(w)\right] = U(CE) = U(E(w) + CE - E(w))$$
$$\approx U(E(w)) + (CE - E(w))U^{'}(E(w)).$$

Comparing the expressions for $E(U(w))$ we have,

$$E(w) - CE = -\frac{1}{2}\frac{U^{''}(E(w))}{U^{'}(E(w))}Var(w) \Rightarrow \text{Risk Premium} = \frac{1}{2}\left[-\frac{U^{''}(E(w))}{U^{'}(E(w))}\right]Var(w).$$

The absolute-risk-aversion (ARA) due to Arrow and Pratt is given by

$$A(w) := -\frac{U^{''}(w)}{U^{'}(w)}.$$

Finally we get the relative-risk-aversion (RRA) as,

$$R(w) := wA(w) = -w\left[\frac{U^{''}(w)}{U^{'}(w)}\right].$$

For a rational investor, we have already seen that $U^{'}(w) > 0$ and $U^{''}(w) < 0$. This gives us that $A^{'}(w) < 0$ and $R^{'}(w) = 0$, i.e., rational investors exhibit decreasing ARA and constant RRA, respectively.

We conclude with some example of common utility functions, as enumerated:

(1) Quadratic Utility: $U(w) = a_1 w - a_2 w^2, a_1 > 0, a_2 > 0, w < \dfrac{a_1}{2a_2}$. Hence
$U'(w) = a_1 - 2a_2 w > 0, U''(w) = -2a_2 < 0, A'(w) > 0$ and $R'(w) > 0$.

(2) Exponential Utility: $U(w) = -e^{-b_1 w}, b_1 > 0$. Hence $U'(w) = b_1 e^{-b_1 w} > 0, U''(w) = -b_1^2 e^{-b_1 w} < 0, A'(w) = 0$ and $R'(w) > 0$.

(3) Logarithmic Utility: $U(w) = \ln(w)$. Hence $U'(w) = \dfrac{1}{w} > 0, U''(w) = -\dfrac{1}{w^2} < 0, A'(w) < 0$ and $R'(w) = 0$.

(4) Power Utility: $U(w) = w^{1-\gamma}, 0 < \gamma < 1$. Hence $U'(w) = (1-\gamma)w^{-\gamma} > 0, U''(w) = -\gamma(1-\gamma)w^{-\gamma-1} < 0, A'(w) < 0$ and $R'(w) = 0$.

Note that the logarithmic utility and the power utility are the ones which satisfy all the four conditions of the risk-averse investor.

Example 5.3.1 Consider an investment opportunity, where an investment of 500, either results in the gain being 100 with probability $\dfrac{1}{2}$, or results in the loss being 100 with probability $\dfrac{1}{2}$. If the utility function for the wealth is $U(w) = \ln(w)$, then determine the certainly equivalent (CE), and the risk premium.

An investment of 500 results in either 600 or 400 each with probability $\dfrac{1}{2}$. Then the expected utility is given by,

$$E(U(w)) = \frac{1}{2} \ln 600 + \frac{1}{2} \ln 400 = 6.1942 = \ln(CE) \Rightarrow CE = e^{6.1942} = 489.8994.$$

Hence, risk premium is $E(w) - CE = 500 - 489.8994 = 10.1006$.

5.4 Exercise

Exercise 5.1 If an investor has the utility function for wealth $U(w) = w^\gamma, (0 < \gamma < 1)$ and is indifferent between receiving 1728 and 512, with equal probabilities, or 1000, with certainty, then determine the utility function explicitly.

Solution: The utilities of the first and second opportunity are $\dfrac{1}{2}(1728)^\gamma + \dfrac{1}{2}(512)^\gamma$ and 1000^γ, respectively. Since the investor is indifferent between these two opportunities, hence,

$$\frac{1}{2}(1728)^\gamma + \frac{1}{2}(512)^\gamma = 1000^\gamma \Rightarrow \gamma = \frac{1}{3}.$$

Hence the explicit utility function is $U(w) = w^{\frac{1}{3}}$.

Exercise 5.2 Consider two investment opportunities A and B over a single period ($t = 0$ to $t = 1$).

Opportunity A: The invested money earns 0% interest rate per period.

Opportunity B: The invested money doubles or halves each with equal probabilities. Now, if an amount of 30 is invested in opportunity A and an amount of 70 is invested in opportunity B with the utility function of wealth being $U(w) = \sqrt{w}$, then determine the expected utility at time $t = 1$.

Solution: In case of the invested money doubling in opportunity B, the total wealth at time $t = 1$ is given by, $30 \times 1 + 70 \times 2 = 170$. On the other hand, in case of the money being halved in opportunity B, the total wealth at time $t = 1$ is given by $30 \times 1 + 70 \times \dfrac{1}{2} = 65$. Hence the expected utility at time $t = 1$ is given by,

$\dfrac{1}{2}\sqrt{170} + \dfrac{1}{2}\sqrt{65} = 10.5503.$

Exercise 5.3 Consider an investment opportunity where an investment of 1000 results in either 1300 or 700, each with probability of $\dfrac{1}{2}$. Now consider a risk-averse, risk neutral, and risk-seeking investor with respective utilities of wealth $U(w) = -w^2$, $U(w) = w$, $U(w) = w^2$. Determine the expected utilities of the three investors.

Solution:

$$E(U_A)(w) = -\frac{1}{2}(1300)^2 - \frac{1}{2}(700)^2 = -1090000,$$

$$E(U_B)(w) = \frac{1}{2} \times 1300 + \frac{1}{2} \times 700 = 1000,$$

$$E(U_C)(w) = \frac{1}{2}(1300)^2 + \frac{1}{2}(700)^2 = 1090000.$$

Exercise 5.4 Consider an investment opportunity, where an investment of 100 gives 110 with probability $\dfrac{2}{3}$ or 90 with probability $\dfrac{1}{3}$. Consider two investors A and B with respective utility of wealth $U_A(w) = (w)^{\frac{1}{3}}$, and $U_B(w) = \ln(w)$. Further the respective CE are CE_A and CE_B. Determine the ratio $\dfrac{CE_A}{CE_B}$.

Solution: Here,

$$CE_A = \left(\frac{2}{3}(110)^{\frac{1}{3}} + \frac{1}{3}(90)^{\frac{1}{3}}\right)^3 = 103.0352,$$

and

$$CE_B = \exp\left\{\frac{2}{3}\ln(110) + \frac{1}{3}\ln(90)\right\} = 102.8828.$$

Therefore,

$$\frac{CE_A}{CE_B} = 1.0015.$$

Exercise 5.5 Consider an investment opportunity, where an investment of 100 gives 105 with probability $\frac{2}{3}$, or 90 with probability $\frac{1}{3}$. Consider two investors A and B with respective utility of wealth $U_A(w) = (w)^{\frac{1}{3}}$, and $U_B(w) = \ln(w)$. Further the respective CE are CE_A and CE_B. Determine the which investor has less risk premium.

Solution: Here,

$$CE_A = \left(\frac{2}{3}(105)^{\frac{1}{3}} + \frac{1}{3}(90)^{\frac{1}{3}}\right)^3 = 99.8283,$$

and

$$CE_B = \exp\left\{\frac{2}{3}\ln(105) + \frac{1}{3}\ln(90)\right\} = 99.7410.$$

Risk premium for investor A is $100 - 99.8283 = 0.1717$, and the risk premium for investor B is $100 - 99.7410 = 0.259$. Clearly, the risk premium for investor B is more than that of investor A.

Non-Mean-Variance Portfolio Theory

6

The discussion on the Markowitz theory and the CAPM was based on the mean-variance framework, wherein the assumption was that the assets follow a normal distribution or that the investors prefer the mean-variance framework. However, empirical analysis of stock market data gives the conclusion that asset returns are not normally distributed. In fact, the distribution is not symmetric and is more peaked, in addition to being leptokurtic, leading to alternative distributions being considered to capture these characteristics of the empirical distribution. Accordingly, in this chapter, we discuss several approaches to non-mean-variance portfolio analysis.

6.1 The Safety First Models

The Markowitz mean-variance approach to the selection of optimal portfolio is based on the utility function, in terms of mean and variance, which means that the outcomes were contingent on the choice of utility functions, and consequently a more objective approach was needed. The safety first models were an alternative to the approach of maximization of expected utility and were motivated by the goal of limiting the possibility of undesirable outcomes in the return of assets. In this section, we elaborate on the three Safety First Criterion. Let r_P denote the return on portfolio P, and let r_D denote the "desired" level of return for an investor, that is, the investor does not wish that r_P falls below r_D. Since r_P is a random variable, the goal is to keep the probability of r_P falling below r_D, to be as small as possible. Let α be the acceptable level of this probability. Consequently, this can mathematically be represented as,

$$P\left(r_P \leq r_D\right) \leq \alpha.$$

S. P. Chakrabarty and A. Kanaujiya, *Mathematical Portfolio Theory and Analysis*,
Compact Textbooks in Mathematics, https://doi.org/10.1007/978-981-19-8544-7_6

6.1.1 Roy's Safety First Criterion

The Roy's Safety First Criterion is to minimize the probability of the portfolio return falling below the desired level, i.e.,

$$\text{Minimize } P\left(r_P < r_D\right).$$

If we now relax the non-normal distribution condition, then,

$$P\left(r_P \leq r_D\right) = P\left(\frac{r_P - \mu_P}{\sigma_P} < \frac{r_D - \mu_P}{\sigma_P}\right) = P\left(z < \frac{r_D - \mu_P}{\sigma_P}\right),$$

where $z \sim N_{0,1}$. Then,

$$\text{Minimize } P\left(r_P < r_D\right) \equiv \text{Minimize } P\left(z < \frac{r_D - \mu_P}{\sigma_P}\right).$$

This is equivalent to,

$$\text{Minimize } \left(\frac{r_D - \mu_P}{\sigma_P}\right) \equiv \text{Maximize } \left(\frac{\mu_P - r_D}{\sigma_P}\right) \equiv \text{Maximize } S_P,$$

taking $r_D = \mu_f$, where S_P is the Sharpe's ratio. From Sharpe's ratio, $\mu_P = \mu_f + S_P \sigma_P$. In summary, the Roy's Safety First Criterion (in case of returns being assumed to be normally distributed, and the desired level of return being the riskfree rate) is equivalent to maximization of the slope S_P, of the straight line $\mu_P = \mu_f + S_P \sigma_P$, in the (σ_P, μ_P)-plane, as shown in Fig. 6.1, with P_4 being the most desired portfolio.

6.1.2 Kataoka's Safety First Criterion

Kataoka's Safety First Criterion sets the objective of choosing the portfolio with the desired return r_D, to be as high as possible, among all portfolios, when the probability of the return of the portfolio not exceeding r_D is less than or equal to α. Mathematically, this is given by,

$$\text{Maximize } r_D \text{ such that } P(r_P < r_D) \leq \alpha.$$

Again, if we relax the non-normal distribution condition, then,

$$P(r_P < r_D) \leq \alpha \iff \frac{r_P - \mu_P}{\sigma_P} \leq -z_\alpha \iff r_P \leq \mu_P - z_\alpha \sigma_P,$$

where z_α is such that $P(z > z_\alpha) = \alpha$. Now the condition of maximization of r_D reduces to

$$r_D = \mu_P - z_\alpha \sigma_P.$$

Geometrically, the Kataoka's Safety First Criterion (in case of the returns being assumed to be normally distributed) is equivalent to the maximization of the intercept r_D of the line $\mu_P = r_D + z_\alpha \sigma_P$ in the (σ_P, μ_P)-plane, as shown in Fig. 6.2, with P_4 being the most desired portfolio.

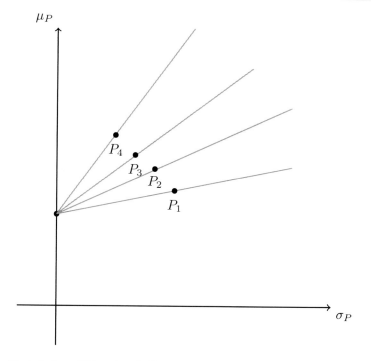

Fig. 6.1 Maximization of Sharpe's ratio S_P

6.1.3   Telser's Safety First Criterion

Telser's Safety First Criterion is based on the premise that investors want to maximize their expected return, subject to the condition that the probability of the return of the portfolio being less than or equal to desired return r_D, does not exceed α. More specifically, this is given by,

$$\text{Maximize } \mu_P \text{ such that } P(r_P \leq r_D) \leq \alpha.$$

If we relax the non-normal distribution condition, then,

$$P(r_P \leq r_D) \leq \alpha \iff \left(\frac{r_D - \mu_P}{\sigma_P}\right) = -z_\alpha \iff r_D \leq \mu_P - z_\alpha \sigma_P$$

$$\iff \mu_P \geq r_D + z_\alpha \sigma_P.$$

Thus the Telser's Safety First Criterion reduces to,

$$\text{Maximize } \mu_P \text{ such that } \mu_P \geq r_D + z_\alpha \sigma_P.$$

Geometrically, the Telser's Safety First Criterion (in case of the returns being assumed to be normally distributed) reduces to the maximization of the expected return among all the portfolios above the line $\mu_P = r_D + z_\alpha \sigma_P$ (called Telser's Ray) in the (σ_P, μ_P)-plane, as shown in Fig. 6.3, with P_4 being the most desired portfolio.

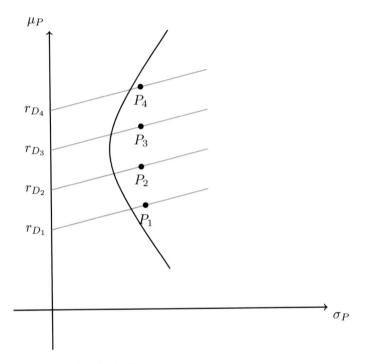

Fig. 6.2 Maximization of r_D for the line $\mu_P = r_D + z_\alpha \sigma_P$

Example 6.1.1 Consider the following table of returns. If $r_D = 4\%$ and the returns are normally distributed, then determine the preferred portfolio(s) using the Roy's Safety First Criterion.

	Portfolio P_1	Portfolio P_2	Portfolio P_3
μ_P	10%	12%	14%
σ_P	9%	11%	13%

The values of $\dfrac{\mu_P - r_D}{\sigma_P}$ for P_1, P_2, and P_3 are 0.6667, 0.7273, and 0.7692, respectively. Since portfolio P_3 has the highest value of $\dfrac{\mu_P - r_P}{\sigma_P}$, hence it is the most preferred one using Roy's Safety First Criterion.

Example 6.1.2 Consider the following table of returns. If the returns are assumed to be normally distributed, with $\alpha = 0.05$, then determine the preferred portfolio(s) using Kataoka's Safety First Criterion.

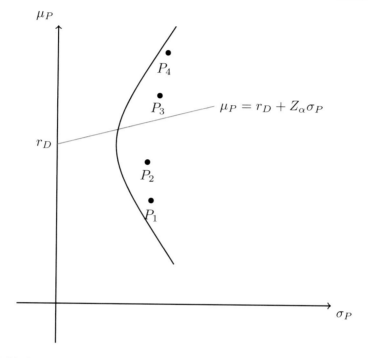

Fig. 6.3 Maximization of μ_P above the line $\mu_P = r_D + Z_\alpha \sigma_P$

	Portfolio P_1	Portfolio P_2	Portfolio P_3
μ_P	8%	10%	13%
σ_P	11%	15%	19%

Since $\alpha = 0.05$, therefore $P(z > 1.645) = 0.05$. Consequently, $z_\alpha = 1.645$. Hence the r_D for each of the three portfolio are,

$$r_{D_1} = 0.08 + (-1.645) \times 0.11 = -0.10095,$$
$$r_{D_2} = 0.10 + (-1.645) \times 0.15 = -0.14675,$$
$$r_{D_3} = 0.13 + (-1.645) \times 0.19 = -0.18255.$$

Since P_1 has the largest value of r_D, among the three, hence P_1 is the most desired portfolio using Kataoka's Safety First Criterion.

Example 6.1.3 Consider three portfolio P_1, P_2, and P_3 with the returns as tabulated below. If the returns are assumed to be normally distributed with, $\alpha = 0.10$ and $r_D = 2\%$, then determine the preferred portfolio(s) using the Telser's Safety First Criterion.

	Portfolio P_1	Portfolio P_2	Portfolio P_3
μ_P	8%	10%	12%
σ_P	5%	8%	7%

Since $\alpha = 0.10$, therefore $P(z > 1.282) = 0.10$. Consequently, $z_\alpha = 1.282$. Hence we check for $\mu_P \geq r_D + z_\alpha \sigma_P$, i.e., $\mu_P - r_D - z_\alpha \sigma_P \geq 0$.

$$\mu_{P_1} - r_D - z_\alpha \sigma_{P_1} = -0.0041,$$
$$\mu_{P_2} - r_D - z_\alpha \sigma_{P_2} = -0.02256,$$
$$\mu_{P_3} - r_D - z_\alpha \sigma_{P_3} = 0.01026.$$

Since P_3 has the largest value among the three (and the only one to satisfy the criterion), hence P_3 is the most desired portfolio using Telser's Safety First Criterion.

6.2 Geometric Mean Return

Let us consider an investment of N periods, with the time points being $i = 0, 1, \ldots, N$. Let r_i denote the return during the period $[i - 1, i)$. Then, an amount invested at $i = 0$ grows by the factor of $(1 + r_1)(1 + r_2) \ldots (1 + r_N)$ at time $i = N$. If $\bar{r}_G$ denotes the geometric mean return, then we have,

$$(1 + \bar{r}_G)^N = \prod_{i=1}^{N}(1 + r_i) \Rightarrow \bar{r}_G = \left(\prod_{i=1}^{N}(1 + r_i)\right)^{\frac{1}{N}} - 1.$$

Now, if we start with wealth w_0, then the wealth at time N is given by

$$w_N = w_0(1 + \bar{r}_G)^N.$$

This gives $E(w_N) = E\left[w_0(1 + \bar{r}_G)^N\right]$, which means that maximization of the terminal wealth is equivalent to the maximization of the geometric mean. If the investor has log utility, then,

$$E[U(w_N)] = E\left[\ln(w_0(1 + \bar{r}_G)^N)\right]$$
$$= E[\ln(w_0) + N\ln(1 + \bar{r}_G)]$$
$$= \ln(w_0) + NE[\ln(1 + \bar{r}_G)].$$

Since w_0 and N are constants, the maximization of expected utility of terminal wealth is equivalent to maximization of expectation of $\ln(1 + \bar{r}_G)$.

Example 6.2.1 A risky asset has five consecutive historical return of 8, 6, 7.5, 5.25, and 6.3%. Then determine the GMR (in percentage)

$$(1 + r_G) = (1.08 \times 1.06 \times 1.075 \times 1.0525 \times 1.063)^{\frac{1}{5}} = 1.066 \Rightarrow r_G = 0.066.$$

Hence the GMR is 6.6%.

6.3   Semi-variance and Semi-deviation

The usage of variance, or equivalently that of standard deviation, as a measure of risk suffers from the shortcoming of penalizing even those scenarios where the portfolio performance is better than a certain threshold, i.e., the cases of high returns. In order to address this, and capture only those returns which are below a threshold, a modified measure of risk, namely semi-variance or equivalently semi-deviation, was formulated. Semi-variance is used to capture the dispersion of those return values, which fall below a certain threshold. Let r^* denote this threshold return for a single period. Then, the semi-variance of the i-th asset is defined as,

$$\overline{\overline{\sigma_i^2}} = E\left[\min(r_i - r^*, 0)\right]^2,$$

which is the average of the square of those returns, which do not exceed the threshold r^*. Analogously, in the continuous setup, the semi-variance is defined as,

$$\overline{\overline{\sigma_i^2}} = \int_{-\infty}^{\infty} \left[\min(r_i - r^*, 0)\right]^2 f(r_i)\mathrm{d}r_i = \int_{-\infty}^{r^*} (r_i - r^*)^2 f(r_i)\mathrm{d}r_i,$$

where $f(r_i)$ is the probability density function of the random variable r_i. Using historical data of returns of the i-th asset, for T time periods, i.e., $r_{i1}, r_{i2}, \ldots, r_{iT}$, the semi-variance can be estimated using,

$$\overline{\overline{\sigma_i^2}} \simeq \frac{1}{T} \sum_{t=1}^{T} \left[\min(r_{it} - r^*, 0)\right]^2.$$

Finally, the semi-deviation is defined as,

$$\overline{\overline{\sigma_i}} = \sqrt{\overline{\overline{\sigma_i^2}}}.$$

Example 6.3.1 If the returns r_i are $10, 13, 7, 12,$ and 11% with $r^* = 10\%$, then determine the semi-variance of the returns (in percentage).

We have the following table. Therefore the semi-variance is given by,

r_i	$r_i - r^*$	$\min(r_i - r^*, 0)$
10	0	0
13	3	0
7	−3	−3
12	2	0
11	1	0

$$\overline{\overline{\sigma_i^2}} = \frac{1}{5}(0 + 0 + 9 + 0 + 0) = 1.8\%.$$

In order to analyze the semi-variance in the $[\sigma, E(r)]$-plane, we consider (for illustrative purpose) the quadratic utility,

$$U(r) = a_1 r - a_2 r^2 \quad \left(r < \frac{a_1}{2a_2}\right).$$

Therefore

$$
\begin{aligned}
E[U(r)] &= a_1 E(r) - a_2 E(r^2) \\
&= a_1 E(r) - a_2 \left[\sigma^2 + (E(r))^2\right] \\
&= a_1 E(r) - a_2 \sigma^2 - a_2 (E(r))^2 \\
&= f_1(\sigma, E(r)).
\end{aligned}
$$

Here $\dfrac{\partial f_1}{\partial E(r)} = a_1 - 2a_2 E(r) > 0$ and $\dfrac{\partial f_1}{\partial \sigma} = -2a_2\sigma < 0$, which means that the expected utility as a function of $E(r)$ and σ is increasing and decreasing, respectively. Now, considering the $[\overline{\overline{\sigma}}, E(r)]$-plane, we propose the following quadratic utility function,

$$U(r) = a_1 r - a_2 \left[\min((r_i - r^*), 0)\right]^2 \quad (a_2 > 0).$$

Therefore,

$$
\begin{aligned}
E[U(r)] &= a_1 E(r) - a_2 E\left[\min((r_i - r^*), 0)\right]^2 \\
&= a_1 E(r) - a_2 \overline{\overline{\sigma}}^2 \\
&= f_2(\overline{\overline{\sigma}}, E(r)).
\end{aligned}
$$

Here $\dfrac{\partial f_2}{\partial E(r)} = a_1 > 0$ and $\dfrac{\partial f_2}{\partial \overline{\overline{\sigma}}} = -2a_2\overline{\overline{\sigma}} < 0$, which shows that the expected utility is an increasing function of $E(r)$ and a decreasing function of $\overline{\overline{\sigma}}$. Now recall that, for a portfolio P,

$$\mu_P = \sum_{i=1}^{n} w_i \mu_i, \quad \sum_{i=1}^{n} w_i = 1.$$

Then the portfolio semi-variance is,

$$\overline{\overline{\sigma_P}}^2 = E\left[\min(r_P - r^*, 0)\right]^2 = E\left[\left(\min\left[\sum_{i=1}^{n} w_i r_i - r^*, 0\right]\right)^2\right].$$

Finally, the CML for the semi-variance is given by,

$$\mu_P = \mu_f + \left[\frac{\mu_m - \mu_f}{\overline{\overline{\sigma}}_m}\right]\overline{\overline{\sigma}}_P,$$

and the corresponding SML is given by,

$$\mu_i = \mu_f + \left[\mu_m - \mu_f\right]\overline{\overline{\beta}}_i,$$

where

$$\overline{\overline{\beta}}_i := \frac{\overline{\overline{\sigma}}_{im}}{\overline{\overline{\sigma}}_m^2} = \frac{E\left[(r_i - r^*)\min(r_m - r^*, 0)\right]}{E\left[\min(r_m - r^*, 0)\right]^2},$$

is what is known as the "downside beta".

6.4 Stochastic Dominance

We consider two portfolios P_1 and P_2, with returns r_{P_1} and r_{P_2}, respectively, and with identical initial investment. The portfolio P_1 is said to dominate P_2 if $r_{P_1} > r_{P_2}$, at the terminal point, for all possible scenarios. However, in practice, this is unlikely to be the case. We discuss in this section, the concepts of the first-, second-, and third-order stochastic dominance. Recall that the cumulative distribution function $F(x)$ is defined as,

$$F(x) = P(X \le x) = \int\limits_{-\infty}^{x} f(t)dt.$$

Stochastic dominance makes use of the probability distribution in its entirety.

6.4.1 First-Order Stochastic Dominance

The first-order stochastic dominance (FSD) is used when the cumulative distribution of return of a portfolio (or asset) P_1 lies above the cumulative distribution of another portfolio (or asset) P_2, in a certain range of return and vice-versa, i.e., P_1 lies below P_2, for certain other range of returns.

Definition 6.4.1 *(FSD) Assuming that the investor has positive marginal utility, i.e., $U'(r) > 0$, portfolio P_1 dominates portfolio P_2 in FSD iff $F_{P_1}(r) \le F_{P_2}(r) \; \forall r$ in the range of return $[\alpha, \beta]$, with $F_{P_1}(r) < F_{P_2}(r)$ for at least one r in the range of return $[\alpha, \beta]$. Here $F_{P_1}(r)$ and $F_{P_2}(r)$ are the cumulative distribution of the returns of portfolio P_1 and portfolio P_2, respectively.*

Theorem 6.4.2 *(FSD Theorem) Let the portfolios P_1 and P_2 have the returns r_{P_1} and r_{P_2}, respectively. For an investor with positive marginal utility, i.e., $U'(r) > 0$ and if portfolio P_1 dominates P_2 in FSD, then the expected utility of P_1 is not less than the expected utility of P_2, i.e., $E\left[U_{P_1}(r)\right] \ge E\left[U_{P_2}(r)\right]$. In other words, portfolio P_1 is preferred to portfolio P_2, in the sense of FSD.*

Proof We recall that the expected utility for P_1 and P_2 is defined for $r \in [\alpha, \beta]$ as,

$$E\left[U_{P_1}(r)\right] = \int\limits_{\alpha}^{\beta} U(r)dF_{P_1}(r) \text{ and } E\left[U_{P_2}(r)\right] = \int\limits_{\alpha}^{\beta} U(r)dF_{P_2}(r),$$

where

$$F_{P_1}(r) = \int\limits_{-\infty}^{r} f_{P_1}(t)dt \text{ and } F_{P_2}(r) = \int\limits_{-\infty}^{r} f_{P_2}(t)dt,$$

are the c.d.f. of returns on P_1 and P_2, respectively. Now,

$$E\left[U_{P_1}(r)\right] - E\left[U_{P_2}(r)\right],$$

$$= \int_\alpha^\beta U(r)\mathrm{d}\left[F_{P_1}(r) - F_{P_2}(r)\right],$$

$$= \left[U(r)\left[F_{P_1}(r) - F_{P_2}(r)\right]\right]_\alpha^\beta - \int_\alpha^\beta U'(r)\left[F_{P_1}(r) - F_{P_2}(r)\right]\mathrm{d}r.$$

Note that $F_{P_1}(\alpha) - F_{P_2}(\alpha) = 0 - 0 = 0$ and $F_{P_1}(\beta) - F_{P_2}(\beta) = 1 - 1 = 0$. Therefore,

$$E\left[U_{P_1}(r)\right] - E\left[U_{P_2}(r)\right] = -\int_\alpha^\beta U'(r)\left[F_{P_1}(r) - F_{P_2}(r)\right] \geq 0,$$

since $U'(r) > 0$ is given, and $F_{P_1}(r) - F_{P_2}(r) \leq 0$, due to FSD, with strict inequality holding for at least one $r \in [\alpha, \beta]$. Hence $E\left[U_{P_1}(r)\right] \geq E\left[U_{P_2}(r)\right]$, i.e., P_1 is preferred to P_2 by the investor. $\qquad\qquad\square$

Corollary 6.4.3 *The following are equivalent:*

(i) *Portfolio P_1 dominates portfolio P_2 in FSD.*
(ii) *$F_{P_1}(r) \leq F_{P_2}(r) \ \forall r \in [\alpha, \beta]$, with strict inequality holding for at least one $r \in [\alpha, \beta]$.*
(iii) *P_1 is preferred to P_2 by an investor with positive marginal utility, i.e., $E\left[U_{P_1}(r)\right] \geq E\left[U_{P_2}(r)\right]$.*

6.4.2 Second-Order Stochastic Dominance

The second-order stochastic dominance (SSD) is used when the cumulative distribution of return of P_1 crossovers the cumulative distribution of return of P_2.

Definition 6.4.4 *(SSD) Assuming that the investor has positive marginal utility, i.e., $U'(r) > 0$ and is risk-averse, i.e., $U''(r) < 0$, portfolio P_1 dominates portfolio P_2 in SSD iff*
$$\int_\alpha^r F_{P_1}(x)dx \leq \int_\alpha^r F_{P_2}(x)dx \ \forall r \ \text{in the range of} \ [\alpha, \beta] \ \text{of returns, with}$$
$$\int_\alpha^r F_{P_1}(x)dx < \int_\alpha^r F_{P_2}(x)dx, \text{for at least one } r \text{ in the range of return } [\alpha, \beta].$$

Theorem 6.4.5 (SSD Theorem) *For an investor with positive marginal utility, i.e., $U'(r) > 0$ and risk-averse attitude, i.e., $U''(r) < 0$, and portfolio P_1 dominates portfolio P_2 in SSD, and expected return from P_1 and P_2 are equal, i.e., $E(r_{P_1}) =$*

$E(r_{P_2})$, then expected utility of P_1 is not less than the expected utility of P_2, i.e., $E\left[U_{P_1}(r)\right] \geq E\left[U_{P_2}(r)\right]$. In other words, the portfolio P_1 is preferred to P_2 in the sense of SSD.

Proof From the proof for FSD,

$$E\left[U_{P_1}(r)\right] - E\left[U_{P_2}(r)\right] = -\int_\alpha^\beta U'(r)\left[F_{P_1}(r) - F_{P_2}(r)\right]dr.$$

Now we consider the term,

$$-\int_\alpha^\beta U'(r)\left[F_{P_1}(r) - F_{P_2}(r)\right]dr$$

$$= \left[-U'(r)\int_\alpha^\beta \left[F_{P_1}(x) - F_{P_2}(x)\right]dx\right]_\alpha^\beta$$

$$+ \int_\alpha^\beta U''(r)\int_\alpha^r \left[F_{P_1}(x) - F_{P_2}(x)\right]dx\,dr$$

$$= -U'(\beta)\int_\alpha^\beta \left[F_{P_1}(x) - F_{P_2}(x)\right]dx$$

$$+ \int_\alpha^\beta U''(r)\int_\alpha^r \left[F_{P_1}(x) - F_{P_2}(x)\right]dx\,dr.$$

Now the integrand in the first term is,

$$\int_\alpha^\beta \left[F_{P_1}(x) - F_{P_2}(x)\right]dx = E(r_{P_1}) - E(r_{P_2}) = 0.$$

Therefore,

$$E\left[U_{P_1}(r)\right] - E\left[U_{P_2}(r)\right] = \int_\alpha^\beta U''(r)\int_\alpha^r \left[F_{P_1}(x) - F_{P_2}(x)\right]dx\,dr \geq 0,$$

since $U''(r) < 0$ and $\int_\alpha^r F_{P_1}(x)dx \leq \int_\alpha^r F_{P_2}(x)dx$, due to SSD, with strict inequality holding for at least one $r \in [\alpha, \beta]$. Hence $E\left[U_{P_1}(r)\right] \geq E\left[U_{P_2}(r)\right]$, i.e., P_1 is preferred to P_2 by the investor. $\qquad\square$

Corollary 6.4.6 *The following are equivalent:*

(i) *Portfolio P_1 dominates portfolio P_2 in SSD.*

(ii) $\int_\alpha^r F_{P_1}(x)dx \le \int_\alpha^r F_{P_2}(x)dx\ \forall r \in [\alpha, \beta]$, *with strict inequality holding for at least one $r \in [\alpha, \beta]$.*

(iii) P_1 *is preferred to P_2, by an investor with positive marginal utility and risk-averse attitude i.e.,* $E\left[U_{P_1}(r)\right] \ge E\left[U_{P_2}(r)\right]$.

6.4.3 Third-Order Stochastic Dominance

The third-order stochastic dominance (TSD) is used when the accumulation of the cumulative distribution of return of P_1 crossovers the accumulation of the cumulative distribution of return of P_2.

Definition 6.4.7 *(TSD) Assuming that the investor have positive marginal utility, i.e., $U'(r) > 0$, is risk-averse, i.e., $U''(r) > 0$ and has decreasing ARA, i.e., $U'''(r) > 0$, portfolio P_1 dominates portfolio P_2 in TSD iff*

$$\int_\alpha^{r_1} \int_\alpha^{r_2} F_{P_1}(x)dxdr_2 \le \int_\alpha^{r_1} \int_\alpha^{r_2} F_{P_2}(x)dxdr_2\ \forall r \text{ in the range of } [\alpha, \beta] \text{ of returns}$$

with

$$\int_\alpha^{r_1} \int_\alpha^{r_2} F_{P_1}(x)dxdr_2 < \int_\alpha^{r_1} \int_\alpha^{r_2} F_{P_2}(x)dxdr_2, \text{ for at least one } r \text{ in the range of return } [\alpha, \beta].$$

Theorem 6.4.8 (TSD Theorem) *For investor with positive marginal utility, i.e., $U'(r) > 0$, risk-averse attitude, i.e., $U''(r) < 0$, decreasing ARA, $U'''(r) > 0$, and portfolio P_1 dominates portfolio P_2 in TSD, and expected return of P_1 is greater than or equal to that of P_2, i.e., $E(r_{P_1}) \ge E(r_{P_2})$, then expected utility of P_1 is not less than the expected utility of P_2, i.e., $E\left[U_{P_1}(r)\right] \ge E\left[U_{P_2}(r)\right]$. In other words, the portfolio P_1 is preferred to P_2 in the sense of TSD.*

Proof From the proof for SSD,

$$E\left[U_{P_1}(r)\right] - E\left[U_{P_2}(r)\right]$$

$$= \int_\alpha^\beta U''(r) \int_\alpha^r \left[F_{P_1}(x) - F_{P_2}(x)\right] dxdr$$

$$= \left[U''(r) \int_\alpha^r \int_\alpha^\beta \left[F_{P_1}(x) - F_{P_2}(x)\right] dxdr \right]_\alpha^\beta$$

$$- \int_\alpha^\beta U'''(r) \int_\alpha^{r_1} \int_\alpha^{r_2} \left[F_{P_1}(x) - F_{P_2}(x)\right] dxdr_1 dr_2.$$

We now consider this term by term. For the first term, $\int_{\alpha}^{\beta} \left[F_{P_1}(x) - F_{P_2}(x)\right] dx \geq 0$,

using $E(r_{P_1}) \geq E(r_{P_2})$ and $U''(r) > 0$, by assumption. Hence the first term is non-negative. For the second term, $\int_{\alpha}^{r_1} \int_{\alpha}^{r_2} \left[F_{P_1}(x) - F_{P_2}(x)\right] dx\, dr_1 dr_2 \leq 0$, using the

TSD condition, and $U'''(r) > 0$, by assumption. Hence the second term is also greater than or equal to zero. Therefore, $E\left[U_{P_1}(r)\right] \geq E\left[U_{P_2}(r)\right]$, i.e., P_1 is preferred to P_2 by the investor.

Example 6.4.9 Consider two companies A and B with the respective returns of r_A and r_B. The values of r_A and r_B along with their corresponding probabilities are tabulated below. Then determine which of the two companies are preferred under the SSD criterion.

r_A	r_B	Probability
5	4	$\frac{1}{3}$
6	6	$\frac{1}{3}$
7	8	$\frac{1}{3}$

The distribution of all possible return is tabulated below.

Since $\int_{\alpha}^{r} F_A(r)dr \leq \int_{\alpha}^{r} F_A(r)dr$ with strict inequality holding for the returns of 4, 5, and 6, hence the return of company A dominates that of company B, in SSD.

Return (r)	$F_A(r)$	$F_B(r)$	$\int_{\alpha}^{r} F_A(r)dr$	$\int_{\alpha}^{r} F_B(r)dr$
4	0	$\frac{1}{3}$	0	$\frac{1}{3}$
5	$\frac{1}{3}$	$\frac{1}{3}$	$\frac{1}{3}$	$\frac{2}{3}$
6	$\frac{2}{3}$	$\frac{2}{3}$	1	$\frac{4}{3}$
7	1	$\frac{2}{3}$	2	2
8	1	1	3	3

Example 6.4.10 Consider two companies A and B with respective returns of r_A and r_B. The values of r_A and r_B along with the corresponding probabilities are tabulated below.

Then determine which of the two companies are preferred under the TSD criterion.

Return	Probability (A)	Probability (B)
4	0.50	0.50
5	0.00	0.00
6	0.00	0.05
7	0.25	0.15
8	0.00	0.00
9	0.00	0.10
10	0.25	0.20

The cumulative distribution of all possible return is tabulated below.

Since $\int_{\alpha}^{r_1} \int_{\alpha}^{r_2} F_A(r) dx dr_2 \leq \int_{\alpha}^{r_1} \int_{\alpha}^{r_2} F_A(r) dx dr_2$ with strict inequality holding for the returns of 6 and 7, hence the return of company A dominates that of company B, in TSD.

Return	$F_A(r)$	$F_B(r)$	$\int_{\alpha}^{r} F_A(r)dr$	$\int_{\alpha}^{r} F_B(r)dr$	$\int_{\alpha}^{r_1}\int_{\alpha}^{r_2} F_A(r)dxdr_2$	$\int_{\alpha}^{r_1}\int_{\alpha}^{r_2} F_B(r)dxdr_2$
4	0.50	0.50	0.50	0.50	0.50	0.50
5	0.50	0.50	1.00	1.00	1.50	1.50
6	0.50	0.55	1.50	1.55	3.00	3.05
7	0.75	0.70	2.25	2.25	5.25	5.30
8	0.75	0.70	3.00	2.95	8.25	8.25
9	0.75	0.80	3.75	3.75	12	12
10	1	1	4.75	4.75	16.75	16.75

Corollary 6.4.11 *The following are equivalent:*

(i) *Portfolio P_1 dominates portfolio P_2 in TSD.*

(ii) $\int_{\alpha}^{r_1} \int_{\alpha}^{r_2} F_{P_1}(x)dxdr_2 \leq \int_{\alpha}^{r_1} \int_{\alpha}^{r_2} F_{P_2}(x)dxdr_2 \ \forall r \in [\alpha, \beta]$, *with the strict inequality holding for at least one $r \in [\alpha, \beta]$.*

(iii) *P_1 is preferred to P_2 by an investor with positive marginal utility, risk-averse attitude and decreasing ARA, i.e., $E\left[U_{P_1}(r)\right] \geq E\left[U_{P_2}(r)\right]$.*

6.5   Portfolio Performance Analysis

1. *Sortino Ratio:* Recall that the Sharpe ratio is given by the excess return over standard deviation of returns. Having recognized that standard deviation penalizes also those returns, which are more than the expected return, we introduced the concept of semi-variance. Accordingly, we now define the counterpart of the

Sharpe ratio (using semi-deviation) to be the Sortino ratio, which is defined as,

$$S_o = \frac{\mu_P - \mu_D}{\bar{\bar{\sigma}}_i},$$

where μ_D is the desired return and $\bar{\bar{\sigma}}_i$ is the semi-deviation.

2. *Maximum Drawdown*: The notion of maximum drawdown determines the extent of how sustained are the losses to a portfolio. Let us consider a time window of $[0, T]$, and a portfolio whose value at time t is given by $P(t), t \in [0, T]$. Then the maximum drawdown is defined as,

$$md(T) = \max_{0 \le t \le T} [m(t) - P(t)],$$

where $m(t) = \max_{0 \le s \le t} P(s)$. Note that, $m(t)$ is the maximum value of the portfolio in the interval $[0, t]$.

6.6 Exercise

Exercise 6.1 Suppose that an investor has invested an amount of 1 million and wishes to have 1.1 million after one year by investing in one of the two following portfolios alternatives.

	Portfolio P_1	Portfolio P_2
μ_P	11%	12%
σ_P	15%	17%

Now, determine, which of the two portfolios P_1 and P_2 should the investor invest in (assuming that the returns are normally distributed). Further, if we designate the chosen portfolio as Q, then determine $P(r_Q < r_D)$.

Solution: Since the investor wishes to have a return of 10% (1 million growing to 1.1 million after one year), so $r_D = 10\%$. Therefore $\frac{\mu_P - r_D}{\sigma_P}$ for P_1 and P_2 are $\frac{1}{15}$ and $\frac{2}{17}$. Since the latter is larger of the two, hence the portfolio P_2 is preferred, using the Roy's Safety First Criterion. Hence P_2 is Q. Therefore $P(r_Q < r_D) = \Phi\left(-\frac{2}{17}\right) = 0.453$.

Exercise 6.2 Consider four portfolios P_1, P_2, P_3, and P_4 with the tabulated returns.

	Portfolio P_1	Portfolio P_2	Portfolio P_3	Portfolio P_4
μ_P	8%	10%	12%	15%
σ_P	5%	8%	7%	8%

If the return are assumed to be normally distributed, with $\alpha = 0.10$, then determine which of these four portfolios achieve(s) the return level of at least 3%, and among them, which is the most desirable, using the Kataoka's Safety First Criterion.

Solution: Since $\alpha = 0.10$, therefore $P(z > 1.282) = 0.10$. Consequently $z_\alpha = 1.282$. Hence the r_D for each of the four portfolios is given by,

$$r_{D_1} = 0.08 + (-1.282) \times 0.05 = 0.0159,$$
$$r_{D_2} = 0.1 + (-1.282) \times 0.08 = -0.00256,$$
$$r_{D_3} = 0.12 + (-1.282) \times 0.07 = 0.03026,$$
$$r_{D_4} = 0.15 + (-1.282) \times 0.08 = 0.04744.$$

Among the four portfolios, portfolios P_3 and P_4 achieve a return level of at least 3%, out of which P_4 (which has the largest value of r_D) is the most desirable using Kataoka's Safety First Criterion.

Exercise 6.3 Consider four portfolios P_1, P_2, P_3, and P_4 with the returns as tabulated below.

	Portfolio P_1	Portfolio P_2	Portfolio P_3	Portfolio P_4
μ_P	10%	12%	15%	12%
σ_P	6%	9%	10%	7%

If the return are assumed to be normally distributed with $\alpha = 0.05$ and $r_D = 2\%$, then determine the preferred portfolio(s) using the Telser's Safety First Criterion.

Solution: Since $\alpha = 0.05$, therefore $P(z > 1.645) = 0.05$. Consequently, $z_\alpha = 1.645$. Hence we check for $\mu_P \geq r_D + z_\alpha \sigma_P$, i.e., $\mu_P - r_D - z_\alpha \sigma_P \geq 0$.

$$\mu_{P_1} - r_D - z_\alpha \sigma_{P_1} = -0.0187,$$
$$\mu_{P_2} - r_D - z_\alpha \sigma_{P_2} = -0.04805,$$
$$\mu_{P_3} - r_D - z_\alpha \sigma_{P_3} = -0.0345,$$
$$\mu_{P_3} - r_D - z_\alpha \sigma_{P_3} = -0.01515.$$

Since none of the four portfolios satisfy the criterion, hence none of them are desirable, using Telser's Safety First Criterion.

Exercise 6.4 A risky asset has five consecutive historical return of 8, 8.2, 7.5, 5.25, and 6.3%. Then determine ratio of the GMR (in percentage) to the average return (in percentage)

Solution: Here,

$$(1+r_G) = (1.08 \times 1.082 \times 1.075 \times 1.0525 \times 1.063)^{\frac{1}{5}} = 1.0704 \Rightarrow r_G = 0.0704.$$

Therefore the GMR is 7.04%. Also,

$$r_{\text{avg}} = \frac{0.08 + 0.082 + 0.075 + 0.0525 + 0.063}{5} = 0.0705.$$

Therefore the average return is 7.05%.

Hence the required ratio is $\dfrac{7.04}{7.05} = 0.999$.

Exercise 6.5 Consider an asset a_i, with $\beta_i = 0.8$ and $\overline{\overline{\beta_i}} = 0.6$, with $E(r_m) = 8\%$ and $\mu_f = 5\%$. If ER and ER_i are the excess return over the riskfree rate in the usual CML and semi-variance CML setup, respectively, then $ER - ER_i$ (in percentage) equals.

Solution: Here,

$$ER - ER_i = \left[E(r_m) - \mu_f\right] \times \left[\beta_i - \overline{\overline{\beta_i}}\right] = (0.08 - 0.05) \times (0.8 - 0.6) = 0.006.$$

Hence the required difference is 0.6%.

Exercise 6.6 Consider two companies A and B, with the respective returns of r_A and r_B. The values of r_A and r_B along with the corresponding probabilities are tabulated below.

r_A	r_B	Probability
9	8	$\frac{1}{4}$
10	10	$\frac{1}{4}$
12	11	$\frac{1}{4}$
13	13	$\frac{1}{4}$

Then determine which of these two companies are preferred under the FSD criterion.

Solution: The cumulative distribution of all possible return is tabulated below.

Return	$F_A(r)$	$F_B(r)$
8	0	$\frac{1}{4}$
9	$\frac{1}{4}$	$\frac{1}{4}$
10	$\frac{2}{4}$	$\frac{2}{4}$
11	$\frac{2}{4}$	$\frac{3}{4}$
12	$\frac{3}{4}$	$\frac{3}{4}$
13	1	1

Since $F_A(r) \leq F_B(r)$ with strict inequality holding for the returns of 8 and 11, hence the return of company A dominates that of company B, in FSD.

Optimal Portfolio Strategies

7

In this chapter, we consider optimization approaches in investment decisions, both in the discrete time and the continuous time setup, making use of the Dynamic Programming Principle and the Hamilton-Jacobi-Bellman equation, respectively.

7.1 Discrete Time Optimization

The Dynamic Programming Principle is a well-established technique of decomposing a complex optimization problem, into a sequence of more tractable, simpler subproblems, and is based on the Principle of Optimality, due to Richard Bellman. The "Principle of Optimality" states that, the optimal solution of a problem is a combination of the optimal solutions to its subproblems. In this section, we use the Dynamic Programming Principle approach, to solve the portfolio optimization problem, in the paradigm of maximization of the expected utility, from the terminal wealth, as well the consumption. This motivates three possible presentations of the objective functional, as enumerated below, for which we denote the wealth and the consumption at time t, by $X(t)$ and $c(t)$, respectively, for the investment horizon of $[0, T]$.

1. Maximization of the expected utility from the terminal wealth, given by,

$$\max E[U_1(X(T))],$$

   where $U_1(\cdot)$ is the utility function for the terminal wealth.
2. Maximization of the expected utility from consumption, given by,

$$\max E\left[\sum_{t=0}^{T} \overline{d}^t U_0(c(t))\right],$$

   where $U_0(\cdot)$ is the utility function for the consumption, and $\overline{d}$ is a discount factor $(0 < \overline{d} < 1)$.

S. P. Chakrabarty and A. Kanaujiya, *Mathematical Portfolio Theory and Analysis*, Compact Textbooks in Mathematics, https://doi.org/10.1007/978-981-19-8544-7_7

3. Maximization of the expected utility from the consumption and the terminal wealth, given by,

$$\max E \left[\sum_{t=0}^{T} \overline{d}^t U_0(c(t)) + U_1(X(T)) \right],$$

where U_0, U_1, and $\overline{d}$ have already been defined.

The optimization problem enumerated above is subject to the constraint dynamics, driven by the structure of and the restrictions on, the portfolio that is being sought to be optimized. We consider the tractable problem of maximization of the expected utility from terminal wealth.

Let the portfolio being considered comprise of n_0 units of a bond and $\{n_i\}_{i=1}^{K}$ units of K stocks, with the price of the bond at time t being $B(t) := S_0(t)$ (for notional brevity) and the price of the i-th stock at time t being $S_i(t), i = 1, 2, \ldots, K$. Let $\mathbf{n} = (n_0, n_1, \ldots, n_K)$ be the tuple of all feasible portfolios, and the objective is to determine the optimal $\mathbf{n}$ that results in the expected utility from terminal wealth being maximized. Now, the constraint, called the self-financing condition, needs to be imposed. This means that if an amount of x is available for investment, then,

$$\sum_{i=0}^{K} n_i S_i(t) = x, \text{ at time } t = 0.$$

In case of consumption of $c(t)$ at times $t = 0, 1, 2, \ldots, T - 1$, the self-financing condition becomes,

$$\sum_{i=0}^{K} n_i^{(t)} S_i(t) = x - c(t),$$

where the newly introduced notation $n_i^{(t)}$ denotes the number of units of the i-th asset purchased at time t and held till time $t + 1$. Note that the consumption at time $t = T$, is simply the terminal wealth.

The model for the bond price and the binomial model for stock price dynamics have already been given in Chap. 3 and will be used to model the dynamics of evolution of the value, $X(t)$, of the portfolio.

For the purpose of motivating the relevance of the Dynamic Programming Principle in the discrete time portfolio optimization problem, we first consider a single period optimization, using the method of Lagrange Multipliers, on the lines of mean-variance portfolio optimization, but with the objective being maximization of expected utility from the terminal wealth, instead of minimization of the variance of the portfolio (as was the case in the mean-variance portfolio optimization). The optimization problem for a single period is given by,

$$\sup_{\mathbf{n}} E[U(X(1))] \text{ subject to } \sum_{i=0}^{K} n_i S_i(0) = x.$$

This gives $X(1) = \sum_{i=0}^{K} n_i S_i(1)$. Accordingly, we define the Lagrangian for the problem as,

$$F(\mathbf{n}, \lambda) = E\left[U\left(\sum_{i=0}^{K} n_i S_i(1)\right)\right] + \lambda\left[x - \sum_{i=0}^{K} n_i S_i(0)\right].$$

We differentiate $F(\mathbf{n}, \lambda)$ with respect to each n_i, $i = 0, 1, 2, \ldots, K$ and set it equal to zero, to obtain,

$$E\left[U'(X^*(1))S_i(1)\right] = \lambda S_i(0), \ i = 0, 1, \ldots, K,$$

where $X^*(1)$ is the optimal terminal wealth, and,

$$\sum_{i=0}^{K} n_i S_i(0) = x.$$

These two equations give a system of $K + 2$ equations, in $K + 2$ unknowns, which have to be solved for $\mathbf{n}$ and λ. In order to determine λ, we note that,

$$S_0(1) = S_0(0)(1 + \mu_f),$$

which gives,

$$E\left[U'(X^*(1))S_0(0)(1 + \mu_f)\right] = \lambda S_0(0) \Rightarrow \lambda = E\left[U'(X^*(1))(1 + \mu_f)\right].$$

Thus, for $i = 0, 1, \ldots K$, we get,

$$E\left[U'(X^*(1))S_i(1)\right] = \frac{S_i(0)}{E\left[U'(X^*(1))(1 + \mu_f)\right]},$$

which has to be solved for $\mathbf{n}$, making use of the binomial model, for $S_i(1)$. As is evident, the approach of Lagrange multipliers being extended for a multi-period optimization problem is computationally cumbersome, which leads to the adoption of Dynamic Programming Principle, so as to tackle the problem. In order to set forth the Dynamic Programming Principle, we introduce the following notations:

$$\mathbf{n}(t) = \left(n_0^{(t)}, n_1^{(t)}, \ldots, n_K^{(t)}\right),$$

where $n_i^{(t)}$ is the number of units of the i-th asset purchased at time t and held till time $t + 1$, i.e., in the interval $[t, t + 1)$ for $t = 0, 1, \ldots, T - 1$ and

$$\mathbf{n}_j = \{\mathbf{n}(j), \mathbf{n}(j + 1), \ldots, \mathbf{n}(T - 1)\},$$

which is the vector notation for the sequence of portfolios from time j to time T, i.e., $[j, T)$. Further, let the optimal strategy in the interval $[t, t + 1)$ be denoted by,

$$\mathbf{n}^*(t) = \left(n_0^*(t), n_1^*(t), \ldots, n_K^*(t)\right).$$

The vector,

$$\mathbf{n}_j^* = \left[\mathbf{n}^*(j), \mathbf{n}^*(j + 1), \ldots, \mathbf{n}^*(T - 1)\right]$$

is the sequence of optimal portfolios from time j to time T, i.e., in the interval $[j, T)$. For the problem under consideration, of maximizing the expected utility from terminal wealth, the goal is to determine the optimal sequence of portfolio $\mathbf{n}_0^*$ from time 0 to time T, i.e., $[0, T)$, using the Dynamic Programming Principle as elaborated below:

Let us consider the problem of determining $\mathbf{n}^*(t)$, in the time interval $[t, t + 1)$, with the wealth level at time t being $X(t) = x$. Let the value function be defined by,

$$v(t, x) := \max_{\mathbf{n}(t)} E_{t,x} \left[U_1(X^{\mathbf{n}_{t+1}^*}(T)) \right].$$

This expression manifests in the expected utility $U_1(\cdot)$ of the terminal wealth $X(T)$, with the optimal portfolio $\mathbf{n}_{t+1}^*$ from time $t + 1$ to T already determined, to be maximized over all feasible portfolio $\mathbf{n}(t)$, held on the interval $[t, t + 1)$, with an available amount of x. This can now be expressed as the Dynamic Programming Principle in the form the following theorem

Theorem 7.1.1 (Dynamic Programming Principle) *Let the value function be defined as,*

$$v(t, x) := \sup_{\mathbf{n}(t)} E_{t,x} \left[v(t + 1, X^{\mathbf{n}(t)}(t + 1)) \right],$$

where $\mathbf{n}(t)$ is an admissible portfolio strategy for $[t, t + 1)$, subject to $v(T, x) = U_1(T, x)$.

Here x is used as a generic variable for the wealth level at the time t given in the argument of v. It is easily observed that the Dynamic Programming Principle for a multi-period setup, reduces to the problem of a series of single period optimization problems, which is now easily handled, without resorting to the approach of Lagrange multipliers.

We now illustrate this, with two examples, for the two utility function identified (in Chap. 5) for risk-averse investors, namely the log utility and the power utility.

Example 7.1.2 Let us consider a portfolio comprising of a stock and a bond. If the investor has the utility $U_1(x) = \log(x)$, then determine the optimal investment strategy using the DPP.

We now apply the Dynamic Programming Principle, starting with $v(T, x) = U_1(x) = \log(x)$. By the backward approach of the Dynamic Programming Principle, we will determine $v(T - 1, x)$. Accordingly at time $t = T - 1$, we let the wealth be $X(T - 1) = x$, and the stock price be $S(T - 1) = s$. Let the number of stocks invested in, by the investor, be n. This means that an amount of ns is invested in the stock and the remaining amount of $x - ns$ is invested in the bond, both at time $t = T - 1$. By the binomial model, the value of the stock at time $T - 1$, which is s, can either become su with probability p, or sd with probability $q = 1 - p$, at time T. Note that, here, we have the up and down factors of u and d, respectively, instead of $(1 + u)$ and $(1 + d)$, used in Chap. 3, for brevity. At the same time, the amount of $x - ns$, invested in the bond at time $T - 1$, grows to $(x - ns)(1 + \mu_f)$, at time T, with

μ_f being the riskfree rate for a single period. This means that $X(T)$ can either take the value of $nsu + (x - ns)(1 + \mu_f)$, with probability p, or $nsd + (x - ns)(1 + \mu_f)$, with probability q. Applying the Dynamic Programming Principle between times $T - 1$ and T, we obtain,

$$
\begin{aligned}
v(T - 1, x) &= \max_n E_{T-1,x}\left[\log(X(T))\right] \\
&= \max_n \left[p \times \log(nsu + (x - ns)(1 + \mu_f)) + q \times \log(nsd + (x - ns)(1 + \mu_f))\right] \\
&= \max_n \left[p \times \log(ns\overline{u} + x(1 + \mu_f)) + q \times \log(ns\overline{d} + x(1 + \mu_f))\right],
\end{aligned}
$$

where $\overline{u} = u - (1 + \mu_f)$ and $\overline{d} = d - (1 + \mu_f)$. Now differentiating the argument, with respect to n and setting equal to zero, we obtain,

$$
w_1^* := \frac{n^* s}{x} = \frac{(1 + \mu_f)[(1 + \mu_f) - pu - qd]}{\overline{u}\,\overline{d}},
$$

where w_1^* is the optimal weight for the stock investment at time $T - 1$ and held upto time T, which is independent of both $X(T - 1) = x$ and $S(T - 1) = s$.

Consequently, substituting the value of n^*, we obtain,

$$
\begin{aligned}
v(T - 1, x) &= \log(x) + p\log\left[(1 + \mu_f)p\left(\frac{\overline{d} - \overline{u}}{\overline{d}}\right)\right] + q\log\left[(1 + \mu_f)q\left(\frac{\overline{u} - \overline{d}}{\overline{u}}\right)\right] \\
&= \log(x) + C_l^{(T-1)},
\end{aligned}
$$

where the constant $C_l^{(T-1)}$ depends only on the model parameters, namely p, q, u, d, and μ_f. In the next step we take,

$$
v(T - 2, x) = \max_n E_{T-2,x}\left[\log(X(T - 1))\right],
$$

with $v(T - 1, x) = \log(x) + C_l^{(T-1)}$. A similar argument will again result in the same n^* in the interval $[T - 2, T - 1)$, and consequently $v(T - 2, x) = \log(x) + C_l^{(T-2)}$, for some parameter dependent only constant $C_l^{(T-2)}$. Thus, the optimal portfolio expression is identical for each subinterval for an investor, with log utility.

Example 7.1.3 In this example, all the assumption and the question are the same as given in Example 7.1.2, except the utility function of $U_1(x) = x^{1-\gamma}$, with $0 < \gamma < 1$.

In this case, applying the Dynamic Programming Principle, between times $T - 1$ and T, we obtain,

$$
\begin{aligned}
v(T - 1, x) &= \max_n E_{T-1,x}\left[(X(T))^{1-\gamma}\right] \\
&= \max_n \left[p \times (nsu + (x - ns)(1 + \mu_f))^{1-\gamma} + q \times (nsd + (x - ns)(1 + \mu_f))^{1-\gamma}\right] \\
&= \max_n \left[p \times (ns\overline{u} + x(1 + \mu_f))^{1-\gamma} + q \times (ns\overline{d} + x(1 + \mu_f))^{1-\gamma}\right],
\end{aligned}
$$

where $\bar{u} = u - (1 + \mu_f)$ and $\bar{d} = d - (1 + \mu_f)$. Now differentiating the argument with respect to n, and setting equal to zero, we obtain,

$$w_1^* = \frac{n^* s}{x} = \frac{(1 + \mu_f)(\kappa - 1)}{\bar{d} - \kappa \bar{u}}, \kappa = \left(-\frac{q\bar{d}}{p\bar{u}}\right)^{\frac{1}{\gamma}},$$

where w_1^* is the optimal weight for the stock investment at time $T - 1$ and held up to time T, which is independent of both $X(T - 1) = x$ and $S(T - 1) = s$. Consequently, substituting the value of n^*, we obtain,

$$v(T - 1, x) = p \left[\bar{u}x \left(\frac{(1 + \mu_f)(\kappa - 1)}{\bar{d} - \kappa \bar{u}}\right) + x(1 + \mu_f)\right]^{1-\gamma}$$
$$+ q \left[\bar{d}x \left(\frac{(1 + \mu_f)(\kappa - 1)}{\bar{d} - \kappa \bar{u}}\right) + x(1 + \mu_f)\right]^{1-\gamma}$$
$$= C_e^{(T-1)} x^{1-\gamma},$$

where the constant $C_e^{(T-1)}$ depends only on the model parameter, namely $p, q, u, d,$ and μ_f. In the next step, we take,

$$v(T - 2, x) = \max_n E_{T-2,x}\left[(X(T - 1))^{1-\gamma}\right],$$

with $v(T - 1, x) = C_e^{(T-1)} x^{1-\gamma}$. A similar argument again results in the same n^* in the interval $[T - 2, T - 1)$, and consequently $v(T - 2, x) = C_e^{(T-2)} x^{1-\gamma}$, for some parameter dependent only on constant $C_e^{(T-2)}$. Thus the optimal portfolio expression is identical for each sub interval for an investor, with power utility.

Example 7.1.4 Consider a portfolio of a stock and a bond with parameter $S(0) = 100, u = 1.1, d = 0.8, \mu_f = 5\%, p = \frac{2}{3}, q = \frac{1}{3}$. Determine the optimal investment strategy after two time periods, on an investment of 500, with the utility functions being:

(i) $U_1(x) = \ln(x)$.
(ii) $U_1(x) = \sqrt{x}$

(i) The optimal weight w_1^* does not depend on the x and s and is given by,

$$w_1^* := \frac{n^* s}{x} = \frac{(1 + \mu_f)((1 + \mu_f) - pu - qd)}{\bar{u}\bar{d}},$$

Now, $\bar{u} = u - (1 + \mu_f) = 1.1 - (1 + 0.05) = 0.05$ and $\bar{d} = d - (1 + \mu_f) = 0.8 - (1 + 0.05) = -0.25$. Hence,

$$w_1^* = \frac{(1 + 0.05)((1 + 0.05) - \frac{2.2}{3} - \frac{0.8}{3})}{0.05 \times (-0.25)} = -4.2.$$

(ii) Given $U_1(x) = \sqrt{x} = x^{1-\frac{1}{2}} \Rightarrow \gamma = \frac{1}{2}$. The optimal weight w_1^* does not depend on the x and s and is given by,

$$w_1^* = \frac{n^* s}{x} = \frac{(1 + \mu_f)(\kappa - 1)}{\overline{d} - \kappa \overline{u}}, \kappa = \left(-\frac{q\overline{d}}{p\overline{u}}\right)^{\frac{1}{\gamma}}.$$

Now, $\overline{u} = u - (1 + \mu_f) = 1.1 - (1 + 0.05) = 0.05$, $\overline{d} = d - (1 + \mu_f) = 0.8 - (1 + 0.05) = -0.25$ and $\kappa = 6.25$. Hence,

$$w_1^* = \frac{(1 + 0.05)(6.25 - 1)}{-0.25 - 6.25 \times 0.05} = -9.8.$$

7.2 Continuous Time Optimization

The Dynamic Programming Principle introduced for multi-period portfolio optimization can be extended for the continuous time setup via what is known as the Hamilton-Jacobi-Bellman (HJB) partial differential equation (PDE), which has established itself beyond just the realm of portfolio optimization.

We consider the HJB PDE approach for the portfolio optimization problem in the paradigm of maximization of the expected utility from the terminal wealth (with the inclusion of consumption being addressed in Sect. 7.3).

We consider a simple portfolio of one bond and one stock. The model for the bond price dynamics and the gBm model for stock prices have already been given in Chap. 3 and will be used to model the dynamics of evolution of the value of $X(t)$ of the portfolio. The bond price and the stock price $B(t)$ and $S(t)$, respectively, at time t in the continuous time setup follow the equations,

$$dB(t) = \mu_f B(t)dt,$$

and

$$dS(t) = \mu S(t)dt + \sigma S(t)dW(t).$$

Let the amount of $X(t)$ be available for investment at time t (with $X(0) = x > 0$), with an amount of $m(t)$ being invested in the stock and the remaining amount of $X(t) - m(t)$ being invested in the bond, at time t. We now drop the argument t for brevity.

Then we can purchase $\frac{m}{S}$ number of stocks, and $\frac{X - m}{B}$ number of bonds, at time t. We now consider a change in the portfolio value, $dX(t)$ in a small time interval dt. Then,

$$dX(t) = \frac{m}{S}dS + \frac{X - m}{B}dB$$

$$= m\left(\frac{dS}{S}\right) + (X - m)\frac{dB}{B}$$

$$= m(\mu dt + \sigma dW(t)) + (X - m)\mu_f dt$$

$$= [\mu_f X + m(\mu - \mu_f)]dt + m\sigma dW(t).$$

Given an admissible portfolio or feasible portfolio $m(t)$, we define,

$$\mathcal{G}(t, x; m) = E_{t,x}[U_1(X^m(T))].$$

Then, the objective of maximization of the expected utility of terminal wealth manifests into the determination of the value function, defined as,

$$v(t, x) := \sup_m E_{t,x}[U_1(X^m(T))] = \sup_m \mathcal{G}(t, x; m),$$

with $m(t) = m(t, X^m(t))$ being considered as a feedback type control. We now introduce the following proposition as a prelude to a "sketch" of the proof of HJB PDE (whose formal derivation is beyond the scope of this book).

Proposition 7.2.1 *The function $\mathcal{G}(t, x; m)$, where $m(t) = m(t, X^m(t))$ (and assuming the existence of $\dfrac{\partial \mathcal{G}}{\partial t}$, $\dfrac{\partial \mathcal{G}}{\partial x}$, and $\dfrac{\partial^2 \mathcal{G}}{\partial x^2}$ and under several conditions) satisfies the PDE,*

$$\frac{\partial \mathcal{G}}{\partial t} + \frac{\partial \mathcal{G}}{\partial x}\left(\mu_f X + m(\mu - \mu_f)\right) + \frac{1}{2}\frac{\partial^2 \mathcal{G}}{\partial x^2}m^2\sigma^2 = 0.$$

Proof A "sketch" of the proof goes as follows.
Application of the Ito's Lemma on $\mathcal{G}(t, x; m) \equiv \mathcal{G}(t, x^m)$ gives,

$$d\mathcal{G} = \frac{\partial \mathcal{G}}{\partial t}dt + \frac{\partial \mathcal{G}}{\partial x}dX + \frac{1}{2}\frac{\partial^2 \mathcal{G}}{\partial x^2}(dX)^2 + \frac{\partial^2 \mathcal{G}}{\partial x\partial t}(dX)(dt) + \frac{1}{2}\frac{\partial^2 \mathcal{G}}{\partial t^2}(dt)^2 + \text{Higher Order Terms}.$$

Using the dynamics of $dX(t)$ and ignoring terms $O((dt)^{3/2})$ and $O((dt)^2)$, along with $(dW)^2 \cong dt$ and $(dW) \cong \sqrt{dt}$, we obtain,

$$d\mathcal{G} = \left[\frac{\partial \mathcal{G}}{\partial t} + \frac{\partial \mathcal{G}}{\partial x}\left(\mu_f X + m(\mu - \mu_f)\right) + \frac{1}{2}\frac{\partial^2 \mathcal{G}}{\partial x^2}\left(m^2\sigma^2\right)\right]dt + \frac{\partial \mathcal{G}}{\partial x}m\sigma dW(t).$$

Now, integrating $\mathcal{G}$ with the initial condition of $X(t) = x$, from t to T, we obtain

$$\mathcal{G}(T, X^m(T))$$
$$= U_1(X^m(T))$$
$$= \mathcal{G}(t, x) + \int_t^T \left[\frac{\partial \mathcal{G}}{\partial y} + \frac{\partial \mathcal{G}}{\partial x}\left(\mu_f X + m(\mu - \mu_f)\right) + \frac{1}{2}\frac{\partial^2 \mathcal{G}}{\partial x^2}\left(m^2\sigma^2\right)\right]dy + \int_t^T \frac{\partial \mathcal{G}}{\partial x}m\sigma dW.$$

Taking conditional expectation on both the sides and rearranging,

$$\mathcal{G}(t, x) = E_{t,x}[U_1(X^m(T))] - E_{t,x}\left[\int_t^T \left[\frac{\partial \mathcal{G}}{\partial y} + \frac{\partial \mathcal{G}}{\partial x}\left(\mu_f X + m(\mu - \mu_f)\right) + \frac{1}{2}\frac{\partial^2 \mathcal{G}}{\partial x^2}\left(m^2\sigma^2\right)\right]dy\right].$$

Note that in this setup we assumed that the expectation can be brought under integral sign in the last term, and the integral with dW is zero ($dW \sim N(0, dt)$, since $E(dW) = 0$). Also recall that,

$$\mathcal{G}(t, x) = E_{t,x}[U_1(X^m(T))].$$

This results in,

$$E_{t,x}\left[\int_t^T\left[\frac{\partial \mathcal{G}}{\partial y} + \frac{\partial \mathcal{G}}{\partial x}\left(\mu_f X + m(\mu - \mu_f)\right) + \frac{1}{2}\frac{\partial^2 \mathcal{G}}{\partial x^2}(m^2\sigma^2)\right]dy\right] = 0.$$

Assuming the condition that the argument of the expectation is zero, followed by the integrand being zero, results in the equation stated in the proposition.

Theorem 7.2.2 *Recall the value function* $v(t,x) := \sup_m \mathcal{G}(t,x;m)$, *where* $\mathcal{G}$ *has already been defined. Then taking the supremum of the PDE in the above Proposition, we get the HJB PDE as,*

$$\frac{\partial v}{\partial t} + \sup_m\left[(\mu - \mu_f)m\frac{\partial v}{\partial x} + \frac{1}{2}m^2\sigma^2\frac{\partial^2 v}{\partial x^2}\right] + \mu_f x\frac{\partial v}{\partial x} = 0.$$

Note that if the optimal investment in the stock $m^*(t) = m^*(t, X^m(t))$ is of the feedback form, then the supremum can be obtained by differentiating the argument, with respect to m, and setting it equal to zero, to obtain,

$$(\mu - \mu_f)\frac{\partial v}{\partial x} + m\sigma^2\frac{\partial^2 v}{\partial x^2} = 0 \Rightarrow m^*(t,x) = -\left(\frac{\mu - \mu_f}{\sigma^2}\right)\frac{\frac{\partial v}{\partial x}}{\frac{\partial^2 v}{\partial x^2}} = -\frac{P}{\sigma}\frac{\frac{\partial v}{\partial x}}{\frac{\partial^2 v}{\partial x^2}}.$$

where $P := \dfrac{\mu - \mu_f}{\sigma}$ is the risk premium (Recall Shape Ratio). Also, the second

derivative of the argument of the supremum is $\sigma^2\dfrac{\partial^2 v}{\partial x^2}$, which is negative, provided v is a convex function (which can be proved). Substituting the optimal $m^*(t,x)$ in the HJB PDE, we get,

$$\frac{\partial v}{\partial t} - \frac{P^2}{2}\frac{\left(\frac{\partial v}{\partial x}\right)^2}{\frac{\partial^2 v}{\partial x^2}} + \mu_f x\frac{\partial v}{\partial x} = 0,\quad v(T,x) = U_1(x).$$

Example 7.2.3 Let us consider a portfolio comprising of a stock and a bond. If the investor has the log utility $U_1(x) = \log(x)$, then determine the optimal investment strategy using the HJB PDE.

Here $U_1(x) = \log(x)$. In order to solve for $v(t,x)$ in the HJB PDE, we assume that $v(t,x) = \log(x) + c_l(T-t)$. Note that this choice is motivated by a combination of the utility function $\log(x)$ and the natural condition $v(T,x) = \log(x)$ being satisfied at $t = T$, leading to the term $T - t$. Then,

$$\frac{\partial v}{\partial t} = -c_l,\quad \frac{\partial v}{\partial x} = \frac{1}{x}\ \text{ and }\ \frac{\partial^2 v}{\partial x^2} = -\frac{1}{x^2}\ \left(\text{Note that }\frac{\partial^2 v}{\partial x^2} < 0\right).$$

Substituting in the HJB PDE, we get $c_l = \mu_f + \dfrac{P^2}{2}$. Therefore,

$$v(t,x) = \log(x) + \left(\mu_f + \frac{P^2}{2}\right)(T - t).$$

Then $m^*(t, x) = \left(\dfrac{\mu - \mu_f}{\sigma^2}\right) x$. Thus the optimal weight of the stock is $w_1^* = \dfrac{\mu - \mu_f}{\sigma^2}$.

Example 7.2.4 Let us consider a portfolio comprising of a stock and a bond. If the investor has the power utility $U_1(x) = x^{1-\gamma}, 0 < \gamma < 1$, then determine the optimal investment strategy using the HJB PDE.

We assume that $v(t, x) = f(t)x^{1-\gamma}$. Then,

$\dfrac{\partial v}{\partial t} = f'(t)x^{1-\gamma}, \ \dfrac{\partial v}{\partial x} = f(t)(1 - \gamma)x^{-\gamma}$ and $\dfrac{\partial^2 v}{\partial x^2} = -f(t)\gamma(1 - \gamma)x^{-\gamma-1}$ (Note that $\dfrac{\partial^2 v}{\partial x^2} < 0$).

Substituting in the HJB PDE, we get,

$$f'(t) + \left(\frac{P^2}{2\gamma} + \mu_f\right) f(t)(1 - \gamma) = 0. \tag{7.1}$$

and $v(T, x) = U(x) \Rightarrow f(T) = 1$. Now solving this equation, with $f(T) = 1$, we get,

$$f(t) = e^{\left(\frac{P^2}{2\gamma} + \mu_f\right)(1-\gamma)(T-t)}.$$

Therefore,

$$v(t, x) = e^{\left(\frac{P^2}{2\gamma} + \mu_f\right)(1-\gamma)(T-t)} x^{1-\gamma}.$$

Then $m^*(t, x) = \left(\dfrac{\mu - \mu_f}{\sigma^2}\right) \dfrac{x}{\gamma}$. Thus the optimal weight of the stock is $w_1^* = \dfrac{\mu - \mu_f}{\sigma^2 \gamma}$.

7.3   Continuous Time Optimization with Consumption

We now consider the HJB PDE with the objective of maximization of expected utility from terminal wealth, as well as consumption. The continuous consumption process is given by $c(t)$, with the cumulative consumption, from time 0 to t being $c(t) := \int_0^t c(y)dy$. Consequently, the dynamics of the portfolio value $X(t)$, with consumption is given by,

$$dX(t) = [\mu_f X + m(\mu - \mu_f) - c]dt + m\sigma dW(t).$$

The objective function after inclusion of consumption now becomes the determination of the value function,

$$v(t, x) := \sup_{m,c} E_{t,x} \left[\int_0^T U_0(c(y))dy + U_1(X^{m,c}(T)) \right].$$

The resulting HJB PDE, with consumption, is then given by,

$$\frac{\partial v}{\partial t} + \sup_{m,c} \left(((\mu - \mu_f)m - c)\frac{\partial v}{\partial x} + \frac{1}{2}m^2\sigma^2\frac{\partial^2 v}{\partial x^2} + U_0(c) \right) + \mu_f x\frac{\partial v}{\partial x} = 0,$$

with $v(T, x) = U_1(x)$. As before, the optimal investment in the stock is given by,

$$m^*(t, x) = -\frac{P}{\sigma}\frac{\frac{\partial v}{\partial x}}{\frac{\partial^2 v}{\partial x^2}}.$$

In order to determine the optimal consumption, we differentiate the argument with respect to c, and set it equal to zero, to obtain,

$$-\frac{\partial v}{\partial x} + U_0'(c) = 0 \Rightarrow c^*(t) = (U_0')^{-1}\left(\frac{\partial v}{\partial x}(t, X^{m^*,c^*}(t))\right).$$

Again, note that the second derivative of the argument with respect to c is $U_0''(c)$, which is negative for a risk-averse investor.

Example 7.3.1 Let us consider a portfolio comprising of a stock and bond. If the investor has the log utility $U_0(x) = \log(x) = U_1(x)$ for both consumption and terminal wealth, then determine the optimal consumption strategy using the HJB PDE.

For the optimal consumption strategy

$$c^*(t) = \left(U_0'\right)^{-1}\left(\frac{\partial v}{\partial x}(t, X^{m^*,c^*}(t))\right) \text{ and } m^*(t, x) = -\frac{P}{\sigma}\frac{\frac{\partial v}{\partial x}}{\frac{\partial^2 v}{\partial x^2}}.$$

Substituting c^* and m^* in the HJB equation with consumption, we get,

$$\frac{\partial v}{\partial t} + \left(((\mu - \mu_f)m^* - c^*)\frac{\partial v}{\partial x} + \frac{1}{2}(m^*)^2\sigma^2\frac{\partial^2 v}{\partial x^2} + U_0(c^*) \right) + \mu_f x\frac{\partial v}{\partial x} = 0,$$

which reduces to,

$$\frac{\partial v}{\partial t} - \frac{P^2}{2}\frac{\left(\frac{\partial v}{\partial x}\right)^2}{\frac{\partial^2 v}{\partial x^2}} + \mu_f x\frac{\partial v}{\partial x} - c^*\frac{\partial v}{\partial x} + U_0(c^*) = 0. \tag{7.2}$$

Since $U_0(x) = \log(x) \Rightarrow U_0'(x) = \frac{1}{x} \Rightarrow (U_0')^{-1}(x) = \frac{1}{x}$, then the above equation can be written as,

$$\frac{\partial v}{\partial t} - \frac{P^2}{2}\frac{(v_x)^2}{v_{xx}} + \mu_f x\frac{\partial v}{\partial x} - \log(v_x) - 1 = 0,$$

with $v(T, x) = \log(x)$. Now, assuming $v(t, x) = f(t) + g(t)\log(x)$ to be the solution of the above PDE, we get,

$$f'(t) + g'(t)\log(x) - \frac{P^2}{2}g(t) + \mu_f g(t) - \log(g(t)) + \log(x) - 1 = 0,$$

which implies,

$$\left(f'(t) - \frac{P^2}{2} g(t) + \mu_f g(t) - \log(g(t)) - 1 \right) + \log(x)(g'(t) + 1) = 0.$$

which gives,

$$f'(t) - \frac{P^2}{2} g(t) + \mu_f g(t) - \log(g(t)) - 1 = 0 \text{ and } g'(t) + 1 = 0.$$

The terminal condition in this case becomes,

$$v(T, x) = \log(x) \Rightarrow f(T) + g(T) \log(x) = \log(x) \Rightarrow f(T) = 1 \text{ and } g(T) = 1.$$

Now solving $g'(t) + 1 = 0$ with $g(T) = 1$, we get,

$$g(t) = T - t + 1 \Rightarrow v(t, x) = f(t) + (T - t + 1) \log(x).$$

Hence,

$$c^*(t) = \frac{1}{v_x(t, x)} = \frac{x}{T - t + 1}.$$

We have not solved for $f(t)$, since it is not required in the determination of $c^*(t)$.

Example 7.3.2 Let us consider a portfolio comprising of a stock and bond. If the investor has the power utility $U_0(x) = x^{1-\gamma} = U_1(x), 0 < \gamma < 1$ for both consumption and terminal wealth, then determine the optimal consumption strategy using the HJB PDE.

Recall that the HJB PDE in this case is,

$$\frac{\partial v}{\partial t} - \frac{P^2}{2} \frac{\left(\frac{\partial v}{\partial x} \right)^2}{\frac{\partial^2 v}{\partial x^2}} + \mu_f x \frac{\partial v}{\partial x} - c^* \frac{\partial v}{\partial x} + U_0(c^*) = 0.$$

Since $U_0(x) = x^{1-\gamma} \Rightarrow U_0'(x) = \frac{1-\gamma}{x^\gamma} \Rightarrow (U_0')^{-1}(x) = \left(\frac{1-\gamma}{x} \right)^{1/\gamma}$, then the above equation can be written as,

$$\frac{\partial v}{\partial t} - \frac{P^2}{2} \frac{\left(\frac{\partial v}{\partial x} \right)^2}{\frac{\partial^2 v}{\partial x^2}} + \mu_f x \frac{\partial v}{\partial x} - \left(\frac{1-\gamma}{v_x} \right)^{1/\gamma} v_x + \left(\frac{1-\gamma}{v_x} \right)^{(1-\gamma)/\gamma} = 0,$$

with $v(T, x) = x^{1-\gamma}$. Now, assuming $v(t, x) = f(t) + g(t)x^{1-\gamma}$, to be the solution of above the PDE, we get,

$$f'(t) + g'(t)x^{1-\gamma} + \frac{P^2}{2r} g(t)(1-\gamma)x^{1-\gamma} + \mu_f(1-\gamma)g(t)x^{1-\gamma} + \gamma(g(t))^{(\gamma-1)/\gamma}x^{1-\gamma} = 0,$$

which implies,

$$f'(t) + \left(g'(t) + \left(\frac{P^2}{2} + \mu_f \right)(1-\gamma)g(t) + \gamma(g(t))^{(\gamma-1)/(\gamma)} \right) x^{1-\gamma} = 0.$$

This gives,

$$f'(t) = 0 \text{ and } g'(t) + \left(\frac{P^2}{2} + \mu_f\right)(1 - \gamma)g(t) + \gamma(g(t))^{(\gamma-1)/(\gamma)} = 0.$$

The terminal condition, in this case, becomes,

$$v(T, x) = x^{1-\gamma} \Rightarrow f(T) + g(T)x^{1-\gamma} = x^{1-\gamma} \Rightarrow f(T) = 0 \text{ and } g(T) = 1.$$

Now on solving $f'(t) = 0$ with $f(T) = 0$, we get $f(t) = 0$.

Now solving, $g'(t) + \left(\frac{P^2}{2} + \mu_f\right)(1 - \gamma)g(t) + \gamma(g(t))^{(\gamma-1)/\gamma} = 0$ with $g(T) = 1$, we get,

$$g(t) = \left(-\frac{\gamma - e^{-\frac{A\left(t - \frac{\gamma \log(A+\gamma)+AT}{A}\right)}{\gamma}}}{A}\right)^\gamma \times \left(\frac{\gamma - e^{\frac{A\left(t + \frac{\gamma \log(A+\gamma-A\gamma)+AT-AT\gamma}{A(\gamma-1)}\right)(\gamma-1)}{\gamma}}}{(A(\gamma-1))}\right)^\gamma,$$

where

$$A = \left(\frac{P^2}{2} + \mu_f\right).$$

Hence,

$$c^*(t) = \left(\frac{1 - \gamma}{v_x}\right)^{1/\gamma} = \left(\frac{1 - \gamma}{1 + (1 - \gamma)g(t)x^{-\gamma}}\right)^{1/\gamma}.$$

7.4 Exercise

Exercise 7.1 Consider a single period binomial model setup for a stock, with parameters $S(0) = 100$, $u = 1.2$, $d = 0.9$, $p = \frac{2}{3}$, $q = \frac{1}{3}$, and $\mu_f = 5\%$. If an investor starts with an amount of 500, then (using the method of Lagrange multiplier) determine the optimal number of stocks, which maximizes the expected utility of the wealth after one time period, for the following utility functions:

(i) $U_1(x) = -e^{-3x}$,
(ii) $U_1(x) = \sqrt{x}$.

Solution
Let the portfolio being considered comprise of n_0 units of the bond and n_1 units the stock, with price of the bond and the stock at time t being $B(t)$ and $S(t)$, respectively. If an amount of x is available for investment at time $t = 0$, the the self-financing condition reduces to,

$$n_0 B(0) + n_1 S(0) = x,$$

with the optimization problem for a single period being given by,

$$\sup_{n_0, n_1} E[U_1(X(1))],$$

where $X(1) = n_0 B(1) + n_1 S(1) = n_0 B(0)(1 + \mu_f) + n_1 S(1)$. Accordingly, we define the Lagrangian for the problem as,

$$F(n_0, n_1, \lambda) = E\left[U_1\left(n_0 B(0)(1 + \mu_f) + n_1 S(1)\right)\right] + \lambda\left[x - n_0 B(0) - n_1 S(0)\right].$$

Differentiating with respect to n_0 and n_1, and setting equal to zero, we get,

$$p\overline{u} U_1'\left(n_0 B(0)(1 + \mu_f) + n_1 u S(0)\right) + q\overline{d} U_1'\left(n_0 B(0)(1 + \mu_f) + n_1 d S(0)\right) = 0.$$

Since $n_0 B(0) = x - n_1 S(0)$, we get,

$$p\overline{u} U_1'\left(x(1 + \mu_f) + n_1 \overline{u} S(0)\right) + q\overline{d} U_1'\left(x(1 + \mu_f) + n_1 \overline{d} S(0)\right) = 0. \qquad (7.3)$$

(i) Since $U_1(x) = -e^{-3x} \Rightarrow U_1'(x) = 3e^{-3x}$, therefore from Eq. (7.3), we have,

$$p\overline{u} e^{3n_1(\overline{u}-\overline{d})S(0)} + q\overline{d} = 0 \Rightarrow n_1 = \frac{1}{3(\overline{u}-\overline{d})S(0)} \ln\left(-\frac{q\overline{d}}{p\overline{u}}\right).$$

Hence,

$$n_1^* = \frac{1}{3(0.15 + 0.15)100} \ln\left(\frac{1}{2}\right) = -\frac{1}{90} \ln(2) = -0.0077.$$

(ii) Since $U_1(x) = \sqrt{x} \Rightarrow U_1'(x) = \frac{1}{\sqrt{x}}$, therefore from Eq. (7.3), we have,

$$p^2\overline{u}^2\left(x(1+\mu_f) + n_1\overline{d}S(0)\right) = q^2\overline{d}^2\left(x(1+\mu_f) + n_1\overline{u}S(0)\right) \Rightarrow n_1 = \frac{x(1+\mu_f)(p^2\overline{u}^2 - q^2\overline{d}^2)}{(\overline{u}-\overline{d})S(0)}.$$

Hence,

$$n_1^* = \frac{500 \times 1.05 \times 0.15 \times 0.05}{100(0.15 + 0.15)} = 0.13125.$$

Exercise 7.2 Derive the optimal portfolio using Dynamic Programming Principle for $U_1(x) = -e^{-b_1 x}$, $b_1 > 0$.

Solution
We apply the Dynamic Programming Principle, starting with $v(T, x) = U_1(x) = -e^{-b_1 x}$, $b_1 > 0$, in order to determine $v(T - 1, x)$. Accordingly, let $X(T - 1) = x$ and $S(T - 1) = s$, with the number of stocks invested in being n. This means that an amount of ns is invested in the stock and the remaining amount of $x - ns$ being invested in the bond. Applying the Dynamic Programming Principle between times $T - 1$ and T, we obtain,

$$v(T - 1, x) = \max_n E_{T-1,x}\left[-e^{-b_1 X(T)}\right]$$

$$= \max_n\left[-p \times e^{-b_1(nsu+(x-ns)(1+\mu_f))} - q \times e^{-b_1(nsd+(x-ns)(1+\mu_f))}\right]$$

$$= \max_n\left[-p \times e^{-b_1(ns\overline{u}+x(1+\mu_f))} - q \times e^{-b_1(ns\overline{d}+x(1+\mu_f))}\right].$$

Now differentiating the argument, with respect to n, and setting equal to zero, we obtain,

$$p \times e^{-b_1(n s\bar{u} + x(1+\mu_f))}(s\bar{u}) + q \times e^{-b_1(n s\bar{d} + x(1+\mu_f))}(s\bar{d}) = 0 \Rightarrow -b_1 n s (\bar{u} - \bar{d}) = \ln\left(-\frac{q\bar{d}}{p\bar{u}}\right).$$

Therefore,

$$n^* = -\frac{1}{b_1 s (\bar{u} - \bar{d})} \ln\left(-\frac{q\bar{d}}{p\bar{u}}\right).$$

Finally,

$$w_1^* := \frac{n^* s}{x} = -\frac{1}{b_1 x (\bar{u} - \bar{d})} \ln\left(-\frac{q\bar{d}}{p\bar{u}}\right),$$

where w_1^* is the optimal weight for the stock investment at time $T - 1$ and held up to time T, which is independent of $S(T - 1) = s$. Consequently, substituting the value of n^*, we obtain,

$$v(T - 1, x) = -p \times e^{-b_1(n^* s\bar{u} + x(1+\mu_f))} - q \times e^{-b_1(n^* s\bar{d} + x(1+\mu_f))}$$
$$= C^{(T-1)} e^{-b_1 x (1+\mu_f)},$$

where the constant $C^{(T-1)}$ depends only on the model parameters, namely $p, q, u,$ and d. In the next step, we take,

$$v(T - 2, x) = \max_n E_{T-2,x}\left[-e^{-b_1(X(T-1))}\right],$$

with

$$v(T - 1, x) = C^{(T-1)} e^{-b_1 x (1+\mu_f)}.$$

A similar argument will again result in the same n^* in the interval $[T - 2, T - 1)$, and consequently

$$v(T - 2, x) = C^{(T-2)} e^{-b_1 x (1+\mu_f)},$$

for some parameter dependent only, constant $C^{(T-2)}$. Thus the optimal portfolio expression is identical for each subinterval for an investor with exponential utility.

Exercise 7.3 Derive the optimal portfolio using Dynamic Programming Principle for $U_1(x) = a_1 x - a_2 x^2$ with $a_1 > 0, a_2 > 0, x < \frac{a_1}{2a_2}$.

Solution
We apply the Dynamic Programming Principle, starting with $v(T, x) = U_1(x) = a_1 x - a_2 x^2$, in order to determine $v(T - 1, x)$. Accordingly, let $X(T - 1) = x$ and $S(T - 1) = s$, with the number of stocks invested in being n. This means that an amount of ns is invested in the stock with the remaining amount $x - ns$ being

invested in the bond. Applying the Dynamic Programming Principle between times $T - 1$ and T, we obtain,

$$
\begin{aligned}
v(T - 1, x) &= \max_n E_{T-1,x}\left[a_1(X(T)) - a_2(X(T))^2 \right] \\
&= \max_n \Big[pa_1(nsu + (x - ns)(1 + \mu_f)) - pa_2(nsu + (x - ns)(1 + \mu_f))^2 \\
&\quad + qa_1(nsd + (x - ns)(1 + \mu_f)) - qa_2(nsd + (x - ns)(1 + \mu_f))^2 \Big] \\
&= \max_n \Big[pa_1(ns\bar{u} + x(1 + \mu_f)) - pa_2(ns\bar{u} + x(1 + \mu_f))^2 \\
&\quad + qa_1(ns\bar{d} + x(1 + \mu_f)) - qa_2(ns\bar{d} + x(1 + \mu_f))^2 \Big].
\end{aligned}
$$

Now differentiating the argument, with respect to n and setting equal to zero, we obtain,

$$
a_1 s \left(p\bar{u} + q\bar{d} \right) - 2a_2 sp\bar{u}(ns\bar{u} + x(1 + \mu_f)) - 2a_2 sq\bar{d}(ns\bar{d} + x(1 + \mu_f)) = 0
$$

$$
\Rightarrow a_1 \left(p\bar{u} + q\bar{d} \right) - 2a_2 \left(p\bar{u} + q\bar{d} \right) x(1 + \mu_f) - 2a_2 ns \left(p\bar{u}^2 + q\bar{d}^2 \right) = 0.
$$

Therefore,

$$
n = \frac{\left(p\bar{u} + q\bar{d} \right) \left(a_1 - 2a_2 x(1 + \mu_f) \right)}{2a_2 s \left(p\bar{u}^2 + q\bar{d}^2 \right)}.
$$

Further,

$$
w_1^* := \frac{n^* s}{x} = \frac{\left(p\bar{u} + q\bar{d} \right) \left(a_1 - 2a_2 x(1 + \mu_f) \right)}{2a_2 x \left(p\bar{u}^2 + q\bar{d}^2 \right)},
$$

where w_1^* is the optimal weight for the stock investment at time $T - 1$ and held up to time T, which is independent of $S(T - 1) = s$. Consequently, substituting the value of n^*, we obtain,

$$
\begin{aligned}
v(T - 1, x) &= \big[pa_1(n^* s\bar{u} + x(1 + \mu_f)) - pa_2(n^* s\bar{u} + x(1 + \mu_f))^2 \\
&\quad + qa_1(n^* s\bar{d} + x(1 + \mu_f)) - qa_2(n^* s\bar{d} + x(1 + \mu_f))^2 \big], \\
&= A^{(T-1)} + B^{(T-1)}x + C^{(T-1)}x^2,
\end{aligned}
$$

where the constants $A^{(T-1)}$, $B^{(T-1)}$, and $C^{(T-1)}$ depend only on the model parameters, namely p, q, u, d, and μ_f. In the next step, we take,

$$
v(T - 2, x) = \max_n E_{T-2,x}\left[a_1 X(T - 1) - a_2(X(T - 1))^2 \right],
$$

with $v(T - 1, x) = A^{(T-1)} + B^{(T-1)}x + C^{(T-1)}x^2$. A similar argument will again results in the same n^* in the interval $[T - 2, T - 1)$, and consequently,

$$
v(T - 2, x) = A^{(T-2)} + B^{(T-2)}x + C^{(T-2)}x^2,
$$

for some parameter dependent only, constant $A^{(T-2)}$, $B^{(T-2)}$, and $C^{(T-2)}$. Thus the optimal portfolio expression is identical for each subinterval for an investor with quadratic utility.

Exercise 7.4 Determine one period optimal portfolio for the parameters $u = 1.1$, $d = 0.8$, $\mu = 5\%$, $p = \dfrac{1}{3}$, $q = \dfrac{2}{3}$, using Dynamic Programming Principle in case of $U_1(x) = \log(x)$.

Solution
From Example 7.1.2, the optimal weight for the stock investment at time $T - 1$ and held up to time T is given by,

$$w_1^* = \frac{(1 + \mu_f)\left[(1 + \mu_f) - pu - qd\right]}{\overline{u}\,\overline{d}}.$$

Accordingly, since $u = 1.1$, $d = 0.8$, and $\mu_f = 0.05$, so we have,

$$\overline{u} = u - (1 + \mu_f) = 1.1 - 1.05 = 0.05 \text{ and } \overline{d} = d - (1 + \mu_f) = 0.8 - 1.05 = -0.25.$$

Therefore,

$$w_1^* = \frac{1.05 \times (1.05 - \frac{1.1}{3} - \frac{1.6}{3})}{-0.05 \times 0.25} = -12.6.$$

Exercise 7.5 Solve the HJB PDE for an investor with $U_1(x) = -e^{-b_1 x}$ for $b_1 > 0$.

Solution
Here $U_1(x) = -e^{-b_1 x}$, $b_1 > 0$. In order to solve for $v(t, x)$ in the HJB PDE, we assume that $v(t, x) = -f(t)e^{-b_1 x}$. Then,

$$\frac{\partial v}{\partial t} = -f'(t)e^{-b_1 x}, \quad \frac{\partial v}{\partial x} = b_1 f(t)e^{-b_1 x} \text{ and } \frac{\partial^2 v}{\partial x^2} = -b_1^2 f(t)e^{-b_1 x}.$$

Substituting in the HJB PDE we get,

$$-f'(t)e^{-b_1 x} + \frac{p^2}{2} f(t)e^{-b_1 x} + \mu_f x b_1 f(t)e^{-b_1 x} = 0 \Rightarrow -f'(t) + \frac{p^2}{2} f(t) + \mu_f x b_1 f(t) = 0.$$

Therefore,

$$f(t) = C_e e^{\left(\frac{p^2}{2} + \mu_f x b_1\right)t}.$$

Using $v(T, x) = U_1(x) \Rightarrow f(T) = 1$. Now determining the value of c_e with $f(T) = 1$, we get,

$$f(t) = e^{-\left(\frac{p^2}{2} + \mu_f x b_1\right)(T - t)} \Rightarrow v(t, x) = -e^{-\left[\left(\frac{p^2}{2} + \mu_f x b_1\right)(T - t) + b_1 x\right]}.$$

Exercise 7.6 Solve the HJB PDE for an investor with $U_1(x) = a_1 x - a_2 x^2$, $a_1 > 0$, $a_2 > 0$, $x < \dfrac{a_1}{2a_2}$.

Solution
Here $U_1(x) = a_1 x - a_2 x^2$. In order to solve for $v(t, x)$ in the above HJB PDE, we assume that $v(t, x) = A(t) + B(t)x + C(t)x^2$. Then,

$$\frac{\partial v}{\partial t} = A'(t) + B'(t)x + C'(t)x^2, \quad \frac{\partial v}{\partial x} = B(t) + 2C(t)x, \text{ and } \frac{\partial^2 v}{\partial x^2} = 2C(t).$$

Substituting in the HJB PDE we get,

$$A'(t) + B'(t)x + C'(t)x^2 - \frac{P^2}{4C(t)}(B(t) + 2C(t)x)^2 + \mu_f(B(t) + 2C(t)x) = 0.$$

Setting the coefficients of x^0, x^1, and x^2 equal to zero, we get, $A'(t) - \dfrac{P^2 B^2(t)}{4C(t)} +$

$\mu_f B(t) = 0$, $B'(t) - P^2 B(t)C(t) + 2\mu_f C(t) = 0$, and $C'(t) - P^2 C(t) = 0$.
Using $v(T, x) = U_1(x) \Rightarrow A(T) = 0, B(T) = a_1$, and $C(T) = -a_2$, we can
solve for $A(t)$, $B(t)$, and $C(t)$, thereby obtaining the value of $v(t, x)$.

Exercise 7.7 Solve the HJB PDE with consumption for an investor with
$U_0(x) = U_1(x) = -e^{-b_1 x}, b_1 > 0$.

Solution
Since $U_0(x) = -e^{-b_1 x} \Rightarrow U_0'(x) = b_1 e^{-b_1 x} \Rightarrow (U_0')^{-1}(x) = -\dfrac{1}{b_1} \ln\left(\dfrac{x}{b_1}\right)$, then
the HJB PDE with consumption can be written as,

$$\frac{\partial v}{\partial t} - \frac{P^2}{2}\frac{(v_x)^2}{v_{xx}} + \mu_f x \frac{\partial v}{\partial x} + \frac{v_x}{b_1} \ln\left(\frac{v_x}{b_1}\right) - \frac{v_x}{b_1} = 0,$$

with $v(T, x) = \log(x)$. Now, assuming $v(t, x) = f(t) - g(t)e^{-b_1 x}$ as the solution
of the above PDE, we get,

$$f'(t) - g'(t)e^{-b_1 x} + \frac{P^2}{2}g(t)e^{-b_1 x} + b_1 \mu_f x g(t)e^{-b_1 x} + g(t)e^{-b_1 x} (\ln(g(t)) - b_1 x)$$
$$- g(t)e^{-b_1 x} = 0,$$

which implies

$$e^{-b_1 x}\left(-g'(t) + \frac{P^2}{2}g(t) + b_1 \mu_f x g(t) + g(t)(\ln(g(t)) - b_1 x) - g(t)\right)$$
$$+ f'(t) = 0.$$

Therefore,

$$f'(t) = 0 \text{ and } -g'(t) + \frac{P^2}{2}g(t) + b_1 \mu_f x g(t) + g(t)(\ln(g(t)) - b_1 x) - g(t).$$

The terminal conditions in this case will become,

$$v(T, x) = -b_1 e^{-b_1 x} \Rightarrow f(T) - g(T)e^{-b_1 x} = -b_1 e^{-b_1 x}$$
$$\Rightarrow f(T) = 0 \text{ and } g(T) = b_1.$$

Now solving $f'(t) = 0$ with $f(T) = 0$, we get,

$$f(t) = 0 \Rightarrow v(t, x) = -g(t)e^{-b_1 x}.$$

Hence,

$$c^*(t) = -\frac{1}{b_1} \ln\left(\frac{v_x}{b_1}\right) = -\frac{1}{b_1} \ln\left(g(t)e^{-b_1 x}\right),$$

where $g(t)$ is the solution of

$$-g'(t) + \frac{P^2}{2}g(t) + b_1\mu_f x g(t) + g(t)(\ln(g(t)) - b_1 x) - g(t) \text{ with } g(T) = 1.$$

Exercise 7.8 Solve the HJB PDE with consumption for an investor with $U_0(x) = \ln(x)$ and $U_1 = x^{1-\gamma}, 0 < \gamma < 1$.

Solution

Since $U_0(x) = \ln(x) \Rightarrow U_0'(x) = \frac{1}{x} \Rightarrow (U_0')^{-1}(x) = \frac{1}{x}$, then the HJB PDE with consumption can be written as,

$$\frac{\partial v}{\partial t} - \frac{P^2}{2}\frac{(v_x)^2}{v_{xx}} + \mu_f x \frac{\partial v}{\partial x} - \log(v_x) - 1 = 0,$$

with $v(T, x) = x^{1-\gamma}$. Now, assuming $v(t, x) = f(t) + g(t)x^{1-\gamma}$ as solution of the above PDE, we get,

$$f'(t) + g'(t)x^{1-\gamma} + \frac{P^2}{2}g(t)\frac{1-\gamma}{\gamma}x^{1-\gamma} + \mu_f g(t)(1-\gamma)x^{1-\gamma} - \log(g(t)(1-\gamma)x^{-\gamma}) - 1 = 0.$$

Therefore

$$f'(t) - \log(g(t)(1-\gamma)x^{-\gamma}) - 1 = 0,$$

and

$$g'(t) + (1-\gamma)\left(\frac{P^2}{2\gamma} + \mu_f\right)g(t) = 0.$$

The terminal conditions in this case will become,

$$v(T, x) = x^{1-\gamma} \Rightarrow f(T) + g(T)x^{1-\gamma} = x^{1-\gamma} \Rightarrow f(T) = 0 \text{ and } g(T) = 1.$$

Now solving

$$g'(t) + (1-\gamma)\left(\frac{P^2}{2\gamma} + \mu_f\right)g(t) = 0 \text{ with } g(T) = 1,$$

we get,

$$g(t) = e^{(1-\gamma)\left(\frac{P^2}{2\gamma} + \mu_f\right)(T-t)}.$$

Hence,

$$c^*(t) = \frac{1}{v_x} = \frac{1}{g(t)(1-\gamma)x^{-\gamma}} = \frac{1}{1-\gamma}x^{\gamma}e^{-(1-\gamma)\left(\frac{P^2}{2\gamma} + \mu_f\right)(T-t)}.$$

Exercise 7.9 Solve both the HJB PDE and the HJB PDE with consumption, for an investor with $U_0(x) = \ln(x)$, $U_1 = x^{1-\gamma}$, $0 < \gamma < 1$, having the model parameters $S(0) = 100$, $\mu_f = 5\%$, and $\sigma = 15\%$.

Solution

For HJB PDE, from Example 7.2.4, the optimal weight of the stock is,

$$w_1^* = \frac{\mu - \mu_f}{\sigma^2 \gamma} = \frac{\mu - 0.05}{2.25\gamma}.$$

For HJB PDE with consumption, from the previous exercise, the optimal consumption is

$$c^*(t) = \frac{1}{v_x} = \frac{1}{g(t)(1-\gamma)x^{-\gamma}} = \frac{1}{1-\gamma} x^{\gamma} e^{-(1-\gamma)\left(\frac{p^2}{2\gamma} + 0.05\right)(T-t)}.$$

Bond Portfolio Optimization

8

The identification of bonds as a risk-free asset must be viewed in the context of the deterministic or known nature of its return, provided the ownership of the bond is held onto, until the maturity of the bond. Having said so, any decision to liquidate a bond, prior to its maturity, has a ramification on the price of the bond, as a result of the movement of interest rate, subsequent to the purchase of the bond. This, in turn, is a consequence of the volatility in the interest rate, prevailing in the market, which impacts not only the price, but also the re-investment strategies. These have resulted in money managers being involved in the active management of bond portfolios, in terms of both modeling future interest rates and devising immunization strategies, to immunize the bond portfolios from the volatility in interest rates.

8.1 Basics of Interest Rates

We begin with the concepts of spot interest rates and forward interest rates. Accordingly, we consider an investment from the current time of t up to time T. The interest rate prevailing at time t, for the period from t to T, is known as the spot rate, with $h = T - t$ being called the horizon of the investor. If at time 0, the interest rate is decided for the period of t to T $(0 < t < T)$, then this interest rate is called the forward rate from t to T. However, for the discussion under consideration, we will consider the horizon h and the corresponding spot rate at t. If $B(t)$ is the bond price at time t, and $B(T)$ is the face value or the par value of the bond at time T, then the spot rate $\mu_{t,T}$ for the horizon $h = T - t$, is given by the relations, as enumerated below, for each of the following cases:

(i) h is less than or equal to a year:

$$B(T) = B(t)\left[1 + \mu_{t,T}(T - t)\right].$$

(ii) h is more than a year and is an integer:

$$B(T) = B(t)(1 + \mu_{t,T})^{T-t}.$$

(iii) h is more than a year and is a fraction:

$$B(T) = B(t)(1 + \mu_{t,T})^{\widetilde{T-t}}(1 + \mu_{t,T}\tau),$$

where $T - t = \widetilde{T - t} + \tau$, with $\widetilde{T - t}$ being the integer part of $T - t$, and τ being the fractional part of $T - t$.

If $\mu_{t,T}^0$ is the forward rate for interest between time t and T, but decided at time 0, then,

$$(1 + \mu_{0,t})^t(1 + \mu_{t,T}^0)^{T-t} = (1 + \mu_{0,T})^T \Rightarrow \mu_{t,T}^0 = \left[\frac{(1 + \mu_{0,T})^T}{(1 + \mu_{0,t})^t} \right]^{\frac{1}{T-t}} - 1.$$

8.2   Bond Pricing

Recall from Chap. 1, that there are predominantly two kinds of bonds, namely zero-coupon bonds and coupon bonds.

We first consider a zero-coupon bond for the period $[0, T]$, with the price of the bond (at $t = 0$) being $B(0)$, and the face value at maturity $t = T$, being $B(T)$. If $\mu_{0,T}$ is the spot rate for the time window $[0, T]$, then the price is of the bond is given by,

$$B(0) = \frac{B(T)}{(1 + \mu_{0,T})^T}.$$

In order to consider a coupon bond, we take the time points of $t = 1, \ldots, T$, with the coupon payments at these time points being $C(1), \ldots, C(T)$, with $C(T)$ being the sum of the last periodic coupon and the face value $B(T)$. If $\mu_{0,t}$ denotes the spot rate for the period $[0, t]$, then the price of this coupon bond is given by,

$$B(0) = \sum_{t=1}^{T} \frac{C(t)}{(1 + \mu_{0,t})^t}.$$

As an illustration, we consider a particular case, where all the periodic coupons are an identical amount of c, with the face value of $B(T)$, and all the spot rates are identical to μ_f. Then the price of the bond as a function of μ_f becomes,

$$B_{\mu_f}(0) = \sum_{t=1}^{T} \frac{c}{(1 + \mu_f)^t} + \frac{B(T)}{(1 + \mu_f)^T} = \frac{c}{\mu_f}\left[1 - \frac{1}{(1 + \mu_f)^T} \right] + \frac{B(T)}{(1 + \mu_f)^T}.$$

We now make two observations:

(i) If $T = 0$, then $B_{\mu_f}(0) = B(T)$, which means that the bond is trading at par.

(ii) In case of a perpetual bond, i.e., $T \to \infty$, we have $B_{\mu_f(0)} \to \dfrac{c}{\mu_f}$.

Two fundamental concepts in the discourse on interest rates models are the yield to maturity and the horizon rate of return. The yield to maturity or the internal rate of return is the single rate of interest that renders the price of the bond to be the present value of the coupons and the face value. If μ_f^* denotes the yield to maturity, then,

$$B(0) = \sum_{t=1}^{T} \frac{C(t)}{(1 + \mu_f^*)^t}.$$

On the other hand, the horizon rate of return is motivated by the projection about the future value of the bond, on the premise that the investor holds onto the bond for her/his respective investment horizon. We denote the horizon rate of return as μ_h, and if an investor invests an amount of $B(0)$ with the intended investment horizon of h, then the future value FV_h, of the investment, at horizon h, is given by,

$$FV_h = B(0)(1 + \mu_h)^h,$$

or equivalently,

$$\mu_h = \left(\frac{FV_h}{B(0)} \right)^{\frac{1}{h}} - 1.$$

Now, we need to recognize that the investment horizon h may be less than the maturity T, which means that the value of FV_h as projected at time $t = 0$ may change over time, as the interest rate changes. Accordingly, the prediction of FV_h is not straightforward. In order to drive home the point, we consider an identical interest for all lengths of time, at the time of purchase of the bond and denote it by μ^0. Suppose that, soon after the purchase of the bond, the interest rate changes to μ, and the investor sticks to her/his originally intended horizon of h.

The value of bond, resulting from the movement of the interest rate from μ^0 to μ, results in the bond price being $B_\mu(0)$. Then the future value is,

$$FV_h = B_\mu(0)(1 + \mu)^h.$$

Hence, we obtain,

$$\mu_h = \left(\frac{B_\mu(0)}{B(0)} \right)^{\frac{1}{h}} (1 + \mu) - 1.$$

We will revisit this at a later stage when deliberating on immunization conditions.

Example 8.2.1 Consider a coupon bond with the face value of 100 and annual coupons of 8, with the maturity being 10 years. If the annual compounding rate of interest is 6%, then determine the price of the bond.

Here, $B(10) = 100$, $c = 8$, $\mu_f = 0.06$, and $T = 10$. Then,

$$B_{0.06}(0) = \sum_{t=1}^{10} \frac{8}{(1.06)^t} + \frac{100}{(1.06)^{10}} = 114.7202.$$

8.3   Duration

Duration is a critical concept in the paradigm of bond portfolio management, both as a measure of risk associated with a bond, and in the determination of immunization, necessitated by the volatility in the interest rates.

Definition 8.3.1   The Duration D of a bond having price B is defined as,

$$D = \frac{\left(\sum_{t=1}^{T} \frac{tC(t)}{(1+\mu_f)^t} \right)}{B}.$$

Note that here $C(t)$ is the payment received at time t by the bondholder and μ_f is the bond's yield to maturity (we have dropped the superscript $*$, for convenience), or μ_f may be defined as the common interest rate, applicable for all lengths of time.

The immediate natural question pertains to the rationale behind identifying Duration (defined above), as a measure of risk for the bond. This will be explained based on the following result.

Result 8.3.2

$$D = -\frac{1+\mu_f}{B} \frac{dB}{d\mu_f}.$$

Proof Recall that the price of a bond is given by,

$$B = \sum_{t=1}^{T} \frac{C(t)}{(1+\mu_f)^t}.$$

Treating B as a function of μ_f, we differentiate B with μ_f (this makes sense, since the goal is to ascertain the sensitivity of the bond price to the interest rate μ_f). Therefore,

$$\frac{dB}{d\mu_f} = \sum_{t=1}^{T} \frac{(-t)C(t)}{(1+\mu_f)^{t+1}}$$

$$\Rightarrow \frac{dB}{d\mu_f} = -\frac{1}{1+\mu_f} \sum_{t=1}^{T} \frac{tC(t)}{(1+\mu_f)^t}$$

$$\Rightarrow \frac{1}{B} \frac{dB}{d\mu_f} = -\frac{1}{1+\mu_f} \left(\frac{1}{B} \sum_{t=1}^{T} \frac{tC(t)}{(1+\mu_f)^t} \right)$$

$$\Rightarrow \frac{1}{B} \frac{dB}{d\mu_f} = -\frac{1}{1+\mu_f} D$$

$$\therefore D = -\frac{1+\mu_f}{B} \frac{dB}{d\mu_f}.$$

Observations:

(1) Now the above expression can be rewritten as,

$$D = -\frac{\frac{dB}{B}}{\frac{d(1+\mu_f)}{1+\mu_f}}.$$

Thus D gives the percentage decrease (increase) in the bond price B, for percentage increase (decrease) in $1 + \mu_f$ (or equivalently μ_f).

(2) The above expression reiterates that the bond goes up, if the interest rate comes down, and vice versa, since the Duration (by its definition) is always positive.

Definition 8.3.3 The modified duration D_m, is defined as $D_m = \dfrac{D}{1 + \mu_f}$.

Consequently, $\dfrac{dB}{B} = -D_m d\mu_f$.

We now discuss the relational properties of Duration with three factors, namely the constant coupon rate, the yield, and the maturity.

(1) Duration as a function of the constant coupon rate, with the yield, and the maturity remaining constant, can be obtained using following closed form of Duration, that will be derived below. Recall that the Duration is,

$$D = -\frac{1 + \mu_f}{B} \frac{dB}{d\mu_f},$$

where

$$B = \frac{c}{1 + \mu_f} + \frac{c}{(1 + \mu_f)^2} + \cdots + \frac{c}{(1 + \mu_f)^T} + \frac{B(T)}{(1 + \mu_f)^T}$$

$$= c \sum_{t=1}^{T} \frac{1}{(1 + \mu_f)^i} + \frac{B(T)}{(1 + \mu_f)^T}$$

$$= c \frac{\frac{1}{(1+\mu_f)} \left(1 - \frac{1}{(1+\mu_f)^T}\right)}{1 - \frac{1}{1+\mu_f}} + \frac{B(T)}{(1 + \mu_f)^T}$$

$$= \frac{c}{\mu_f} \left[1 - \frac{1}{(1 + \mu_f)^T}\right] + \frac{B(T)}{(1 + \mu_f)^T}.$$

This gives,

$$\frac{B}{B(T)} = \frac{\frac{c}{B(T)}}{\mu_f} \left[1 - \frac{1}{(1 + \mu_f)^T}\right] + \frac{1}{(1 + \mu_f)^T}$$

$$= \frac{1}{\mu_f} \left[\frac{c}{B(T)} \left(1 - \frac{1}{(1 + \mu_f)^T}\right) + \frac{\mu_f}{(1 + \mu_f)^T}\right].$$

Taking log on both sides we get,

$$\log\left(\frac{B}{B(T)}\right) = -\log\mu_f + \log\left[\frac{c}{B(T)}\left(1 - \frac{1}{(1+\mu_f)^T}\right) + \frac{\mu_f}{(1+\mu_f)^T}\right],$$

which upon differentiation with respect to μ_f, gives,

$$\frac{d\log\left(\frac{B}{B(T)}\right)}{d\mu_f} = \frac{d\log B}{d\mu_f} = \frac{1}{B}\frac{dB}{d\mu_f}$$

$$= -\frac{1}{\mu_f} + \frac{\frac{c}{B(T)}T\frac{1}{(1+\mu_f)^{T+1}} + \frac{1}{(1+\mu_f)^T} + \frac{\mu_f}{(1+\mu_f)^{T+1}}(-T)}{\left[\frac{c}{B(T)}\left(1 - \frac{1}{(1+\mu_f)^T}\right) + \frac{\mu_f}{(1+\mu_f)^T}\right]}.$$

Therefore we get,

$$D = -\frac{1+\mu_f}{B}\frac{dB}{d\mu_f} = 1 + \frac{1}{\mu_f} + \frac{T\left(\mu_f - \frac{c}{B(T)}\right) - (1+\mu_f)}{\frac{c}{B(T)}\left((1+\mu_f)^T - 1\right) + \mu_f}. \qquad (8.1)$$

From this expression, we see that if the coupon rate given by $\dfrac{c}{B(T)}$ increases, with the other quantities remaining fixed, then the numerator value decreases, and the denominator value increases, concurrently resulting in the overall decrease in the Duration.

Thus in summary, the Duration decreases (increases) as the coupon rate increases (decreases).

Note that this conclusion is based on the constant coupon and is arrived at by deriving the closed form expression for the Duration.

(2) The relationship between the Duration and the yield also requires the derivation of an expression for $\dfrac{dD}{d\mu_f}$, as follows. We begin with the general expression for Duration (treating $B := B(\mu_f)$), as follows,

$$D = \frac{1}{B(\mu_f)}\sum_{t=1}^{T}\frac{tC(t)}{(1+\mu_f)^t}.$$

Differentiating with respect to μ_f, we obtain,

$$\frac{dD}{d\mu_f} = -\frac{1}{B^2(\mu_f)}\left[\sum_{t=1}^{T}\frac{t^2C(t)}{(1+\mu_f)^{t+1}}\times B(\mu_f) + \sum_{t=1}^{T}\frac{tC(t)}{(1+\mu_f)^t}\frac{dB}{d\mu_f}\right],$$

$$= -\frac{1}{(1+\mu_f)}\left[\frac{\sum_{t=1}^{T}\frac{t^2C(t)}{(1+\mu_f)^t}}{B(\mu_f)} + (1+\mu_f)\frac{\frac{dB(\mu_f)}{d\mu_f}}{B(\mu_f)}\frac{\sum_{t=1}^{T}\frac{tC(t)}{(1+\mu_f)^t}}{B(\mu_f)}\right],$$

$$= -\frac{1}{(1+\mu_f)}\left[\frac{\sum_{t=1}^{T}\frac{t^2C(t)}{(1+\mu_f)^t}}{B(\mu_f)} - D^2\right].$$

We now define,

$$u(t) := \frac{\frac{C(t)}{(1+\mu_f)^t}}{B(\mu_f)}.$$

Then using the fact that $\sum_{t=1}^{T} u(t) = 1$ and $\sum_{t=1}^{T} tu(t) = D$, we get

$$\frac{dD}{d\mu_f} = -\frac{1}{1+\mu_f}\left[\sum_{t=1}^{T} t^2 u(t) - D^2\right]$$

$$= -\frac{1}{1+\mu_f}\left[\sum_{t=1}^{T} t^2 u(t) - 2D^2 + D^2\right]$$

$$= -\frac{1}{1+\mu_f}\left[\sum_{t=1}^{T} t^2 u(t) - 2D\sum_{t=1}^{T} tu(t) + D^2\sum_{t=1}^{T} u(t)\right]$$

$$= -\frac{1}{1+\mu_f}\sum_{t=1}^{T}(t-D)^2 u(t)$$

$$= -\frac{1}{1+\mu_f}v_D,$$

where $v_D := \sum_{t=1}^{T}(t-D)^2 u(t)$ is the variance of times of payments about the Duration D, with the weights being $u(t)$. Since $u(t) > 0$ and $v_D > 0$, therefore $\frac{dD}{d\mu_f} < 0$.

Thus we conclude that the Duration decreases (increases) as the yield increases (decreases).

(3) In case of relationship between Duration and maturity, there does not exist a clear single relationship, and accordingly, we enumerate some of the contingent relationship.

(A) Since for a zero-coupon bond, the Duration is equal to maturity, hence the Duration increases (decreases) as maturity increases (decreases)).

(B) In case of coupon bearing bonds, we refer to relation (8.1), where we observe that as $T \to \infty$, the second expression tends to zero, since the numerator and denominator, have T as affine function and exponential function, respectively. Hence as $T \to \infty$, $D \to 1 + \frac{1}{\mu_f}$.

(C) In case of coupon bearing bonds, with coupon rates greater than or equal to the yield, we get that an increase in maturity results, in an increase in Duration, tending toward $1 + \frac{1}{\mu_f}$.

(D) Finally in case of coupon bearing bonds, with coupon rates less than the yield, it gives that an increase in maturity results in an initial increase in the

Duration, until a maximum is reached, followed by a decrease tending toward
$1 + \dfrac{1}{\mu_f}$.

Example 8.3.4 Consider a coupon bond with the face value of 100 and annual coupons of 8, with the maturity being 10 years. If the annual compounding rate of interest is 6%, then determine the Duration and the modified Duration of the bond.

From the previous Example (8.2.1), the Duration of the bond is given by,

$$D = \frac{1}{114.7202} \left[\sum_{t=1}^{10} \frac{8t}{(1.06)^t} + \frac{100 \times 10}{(1.06)^{10}} \right] = 7.4450.$$

The modified Duration is given by,

$$D_m = \frac{D}{1 + \mu_f} = \frac{7.4450}{1.06} = 7.0236.$$

8.4 Duration for a Bond Portfolio

We now consider a portfolio of bonds. Let the number of bonds in the portfolio be n. The price of the i-th bond is given by $B_i(\mu_f)$, and the number of units of the i-th bond in the portfolio is given by N_i. Then the value of the bond portfolio is given by,

$$P(\mu_f) = \sum_{i=1}^{n} N_i B_i(\mu_f).$$

Differentiating both sides with respect to μ_f, we obtain,

$$\frac{dP(\mu_f)}{d\mu_f} = \sum_{i=1}^{n} N_i \frac{dB_i(\mu_f)}{d\mu_f}.$$

Now, we multiply both sides by $-\dfrac{1 + \mu_f}{P(\mu_f)}$, to obtain,

$$-\frac{1 + \mu_f}{P(\mu_f)} \frac{dP(\mu_f)}{d\mu_f} = \sum_{i=1}^{n} \frac{N_i B_i(\mu_f)}{P(\mu_f)} \left(-\frac{1 + \mu_f}{B_i(\mu_f)} \frac{dB_i(\mu_f)}{d\mu_f} \right).$$

This gives,

$$D_P = \sum_{i=1}^{n} \frac{N_i B_i(\mu_f)}{P(\mu_f)} D_i,$$

where D_P and D_i are the Duration of the bond portfolio, and the i-th bond, respectively. Finally we consider the term $\dfrac{N_i B_i(\mu_f)}{P(\mu_f)}$. Here, the numerator is the amount invested in the i-th bond, and the denominator is the value of the portfolio, resulting in the expression being the weight of the i-th bond in the portfolio and denoted by w_i, i.e.,

$$w_i := \frac{N_i B_i(\mu_f)}{P(\mu_f)} \text{ with } \sum_{i=1}^{n} w_i = 1.$$

Thus, we have,

$$D = \sum_{i=1}^{n} w_i D_i,$$

which means that the Duration of a bond portfolio is the weighted sum of the Duration of the constituent bonds (note that this is similar to the return of a portfolio of risky assets).

Example 8.4.1 If we invest $\dfrac{2}{3}$-rd and $\dfrac{1}{3}$-rd of the total available amount in zero-coupon bond of maturity 2 years and 3 years respectively, then determine the Duration of the resulting portfolio of these two bonds.

Here, $w_1 = \dfrac{2}{3}$, $w_2 = \dfrac{1}{3}$, $D_1 = 2$ and $D_2 = 3$. Therefore,

$$D_P = \frac{2}{3} \times 2 + \frac{1}{3} \times 3 = 2.3333.$$

8.5   Immunization Using Duration

The susceptibility of bond owners to interest rate movements can be viewed from two perspectives. In case of bonds of shorter maturity or bonds with predetermined coupons (not decided by reinvestment), a rise in interest rates results in losses for the bond holder, due to fall in the bond prices. On the upside, the bond owner gains, in the event of the interest rates going down. The second perspective of the risk stems from the fact that in case of long-term bonds, where the coupons are reinvested, a fall in interest rates means that the reinvestment has to be done at a lower rate. The investor then is faced with the question of an appropriate horizon, to hedge against the likely losses resulting from adverse interest movement, and this is achieved through a process known as *immunization*. It turns out that if the horizon of the investor matches the Duration of the bond, then the investor is immunized or hedged against the consequences of parallel movement of interest rates.

The immediate question that arises is about the scenario where the horizon of the investor does not match with the Duration of the bonds that the investor intends to purchase and the consequent immunization strategy that has to be adopted. This is where the relation for the Duration of a portfolio comes into the picture, which begets the immunization strategy of constructing a bond portfolio whose Duration matches the horizon of the investor, which will now be derived in the form of the following theorem.

Theorem 8.5.1 *For a term structure which is horizontal (same interest rate for all maturities), a bond of Duration D is immunized against any parallel movement of the horizontal term structure, if the investor's horizon h, equals D.*

Proof Let h be the horizon of the investor. Then

$$B(\mu_f)(1 + \mu_f)^h = FV_h = B(0)(1 + \mu_h)^h.$$

Accordingly,

$$\mu_h = \left(\frac{FV_h}{B(0)}\right)^{\frac{1}{h}} - 1.$$

In order to immunize, we need μ_h to be minimized, which is equivalent to minimization of FV_h, or equivalently we minimize $\ln FV_h$. Note that,

$$\ln FV_h = \ln B(\mu_f) + h \ln(1 + \mu_f).$$

Differentiating $\ln FV_h$ with respect to μ_f, and setting it equal to zero, we obtain,

$$
\begin{aligned}
\frac{\mathrm{d}\ln FV_h}{\mathrm{d}\mu_f} &= \frac{\mathrm{d}}{\mathrm{d}\mu_f} \ln B(\mu_f) + \frac{h}{1 + \mu_f} \\
&= \frac{1}{B(\mu_f)} \frac{\mathrm{d}B(\mu_f)}{\mathrm{d}\mu_f} + \frac{h}{1 + \mu_f} \\
&= -\frac{D}{1 + \mu_f} + \frac{h}{1 + \mu_f} \\
&= \frac{(h - D)}{1 + \mu_f} = 0.
\end{aligned}
$$

Therefore $h = D$, that is, the horizon is equal to the Duration. For minima, we now determine the second derivative of $\ln FV_h$ with respect to μ_f as,

$$
\begin{aligned}
\frac{\mathrm{d}^2 \ln FV_h}{\mathrm{d}\mu_f^2} &= \frac{1}{(1 + \mu_f)^2} \left[-\frac{\mathrm{d}D}{\mathrm{d}\mu_f}(1 + \mu_f) + D - h \right] \\
&= -\frac{1}{1 + \mu_f} \frac{\mathrm{d}D}{\mathrm{d}\mu_f} \\
&= -\frac{1}{(1 + \mu_f)} \left(-\frac{1}{1 + \mu_f} v_D \right) \\
&= \frac{v_D}{(1 + \mu_f)^2} > 0.
\end{aligned}
$$

This completes the second-order condition test.

8.6 Convexity

It has been observed that the value of a bond as a function of interest rates is convex, which motivates the notion of Convexity of the bond itself. So far, we have viewed the susceptibility of bond prices to the interest rates, in terms of Duration. This leads us to the investors' dilemma, as to what would be a prudent choice between bonds, when both have the same yield and identical Duration. The investor may be led to believe that both these bonds bear identical interest rate risk. However, the examination of the Convexity of these two bonds leads to a clear preferential criterion, which is the focus of this section. The notation of Convexity of a curve relates to the rate of change of the slope of the curve and is given by the second derivative.

Definition 8.6.1 The Convexity, C of a bond, whose price is B, is defined as,

$$C = \frac{1}{B} \frac{\mathrm{d}}{\mathrm{d}\mu_f} \left(\frac{\mathrm{d}B}{\mathrm{d}\mu_f} \right) = \frac{1}{B} \frac{\mathrm{d}^2 B}{\mathrm{d}\mu_f^2}.$$

Recall that,

$$\frac{\mathrm{d}B}{\mathrm{d}\mu_f} = \sum_{t=1}^{T} \frac{(-t)C(t)}{(1+\mu_f)^{t+1}}.$$

Accordingly,

$$C = \frac{1}{B} \frac{\mathrm{d}^2 B}{\mathrm{d}\mu_f^2} = \frac{1}{B(1+\mu_f)^2} \sum_{t=1}^{T} \frac{t(t+1)C(t)}{(1+\mu_f)^t}.$$

It can be easily seen that the Convexity of a bond is positive, and has the dimension of years^2. We now reconcile the Convexity of a bond as an indicator of the sensitivity of the bond price, to interest rate movements. Accordingly, we consider a small change $\mathrm{d}B$ (from $B(\mu_f)$ to $B(\mu_f + \mathrm{d}\mu_f)$), resulting from a change $\mathrm{d}\mu_f$ (from μ_f to $\mu_f + \mathrm{d}\mu_f$), in the interest rate. Using the Taylor series expansion, and retaining up to second-order terms, we get,

$$\frac{\mathrm{d}B}{B} = \frac{1}{B} \left[B(\mu_f + \mathrm{d}\mu_f) - B(\mu_f) \right],$$

$$\approx \frac{1}{B} \left[\frac{\mathrm{d}B}{\mathrm{d}\mu_f} \mathrm{d}\mu_f + \frac{1}{2} \frac{\mathrm{d}^2 B}{\mathrm{d}\mu_f^2} (\mathrm{d}\mu_f)^2 \right],$$

$$= -D_m \mathrm{d}\mu_f + \frac{1}{2} C (\mathrm{d}\mu_f)^2.$$

From this relation, it can be surmised that a higher Convexity is desirable for the investor.

8.7 Convexity for a Bond Portfolio

We consider a bond portfolio of n bonds with the price of i-th bond being $B_i(\mu_f)$. If N_i is the number of units of the i-th bond, then the value of the portfolio is given by,

$$P(\mu_f) = \sum_{i=1}^{n} N_i B_i(\mu_f).$$

Taking the second derivative of $P(\mu_f)$, with respect to μ_f, and dividing by $P(\mu_f)$, we obtain,

$$\frac{1}{P(\mu_f)} \frac{d^2 P(\mu_f)}{d\mu_f^2} = \sum_{i=1}^{n} \frac{N_i}{P(\mu_f)} \frac{d^2 B_i(\mu_f)}{d\mu_f^2} = \sum_{i=1}^{n} \left(\frac{N_i B_i(\mu_f)}{P(\mu_f)} \right) \frac{1}{B_i(\mu_f)} \frac{d^2 B_i(\mu_f)}{d\mu_f^2}$$

This implies that,

$$C_P = \sum_{i=1}^{n} w_i C_i.$$

Here, C_P and C_i are the Convexity of the bond portfolio and the i-th bond, respectively, with $w_i = \dfrac{N_i B_i(\mu_f)}{P(\mu_f)}$ being the weight of the i-th bond in the bond portfolio. Thus, we have

$$C_P = \sum_{i=1}^{n} w_i C_i,$$

which (like Duration) gives that the Convexity of a bond portfolio is the weighted sum of the Convexity of the constituent bonds of the portfolio.

We now conclude our discussion on Convexity by establishing its functional relationship with Duration. Accordingly, starting with, $D = -\dfrac{(1 + \mu_f)}{B} \dfrac{dB}{d\mu_f}$, we obtain,

$$-\left(\frac{v_D}{(1 + \mu_f)} \right) = \frac{dD}{d\mu_f} = -\left[\frac{B - (1 + \mu_f)\frac{dB}{d\mu_f}}{B^2} \right] \frac{dB}{d\mu_f} - \frac{(1 + \mu_f)}{B} \frac{d^2 B}{d\mu_f^2}.$$

This upon simplification reduces to,

$$\frac{1}{B}(1 + D)\frac{dB}{d\mu_f} + \left(\frac{1 + \mu_f}{B} \right) \frac{d^2 B}{d\mu_f^2} = \frac{v_D}{1 + \mu_f}.$$

Now substituting, $\dfrac{\frac{dB}{d\mu_f}}{B} = -\dfrac{D}{1 + \mu_f}$ and $\dfrac{\frac{d^2 B}{d\mu_f^2}}{B} = C$, we get,

$$-D(1 + D) + (1 + \mu_f)^2 C = v_D.$$

Therefore, we get,

$$C = \frac{1}{(1 + \mu_f)^2}[v_D + D(1 + D)].$$

This relation reveals that the Convexity depends upon both variance of the times and Duration.

Example 8.7.1 Consider a coupon bond with the face value of 100 and annual coupons of 8, with the maturity being 10 years. If the annual compounding rate of interest is 6%, determine its Convexity.

Form the earlier Example (8.2.1), the Convexity of the bond is given by,

$$C = \frac{1}{114.7270 \times (10.6)^2} \left[\sum_{t=1}^{10} \frac{8t(t+1)}{(1.06)^t} + \frac{100 \times 10 \times 11}{(1.06)^{10}} \right] = 65.1716.$$

8.8 Applications

In this section we consider a couple of applications of the concept of Duration.

1. *The Reddington Problem*: Let us consider the time points $t = 1, \ldots, T$. Let $l(t)$ and $a(t)$, be the liabilities and asset, of an entity at time $t = 1, \ldots, T$. Then the present values of the liabilities and asset are,

$$l = \sum_{t=1}^{T} \frac{l(t)}{(1 + \mu_f)^t} \text{ and } a = \sum_{t=1}^{T} \frac{a(t)}{(1 + \mu_f)^t},$$

respectively. Let the net value be $n_v = a - l$, which is zero initially. The goal is to ensure that the n_v is immune to charges in μ_f, which manifests mathematically into the condition of $\frac{dn_v}{d\mu_f} = 0$. Now,

$$\frac{d}{d\mu_f} \left(\frac{\sum_{t=1}^{T}(a(t) - l(t))}{(1 + \mu_f)^t} \right) = \frac{1}{(1 + \mu_f)} \sum_{t=1}^{T} \frac{t(l(t) - a(t))}{(1 + \mu_f)^t}$$

$$= \frac{1}{1 + \mu_f}(D_l l - D_a a),$$

where $D_l = \left(\sum_{t=1}^{T} \frac{tl(t)}{(1 + \mu_f)^t} \right) \frac{1}{l}$, and $D_a = \left(\sum_{t=1}^{T} \frac{ta(t)}{(1 + \mu_f)^t} \right) \frac{1}{a}$, are the Duration of the liability and the asset, respectively.

$$\therefore \frac{dn_v}{d\mu_f} = 0 \Rightarrow \frac{1}{1 + \mu_f}(D_l l - D_a a) = 0 \Rightarrow \frac{a}{1 + \mu_f}(D_l - D_a) = 0,$$

since $l = a$ initially, which gives $D_l = D_a$, i.e., the Duration of liabilities and asset must be identical.

2. *Duration Matching:* Suppose that we have a bond with value B, of a certain Duration D. The goal is the construction of a portfolio of bonds with the same Duration D. Accordingly, let us consider the constructed bond portfolio with N_1 number of a bond with price B_1 and N_2 number of a bond with price B_2. Since we want to create a replicating bond portfolio, so,

$$N_1 B_1 + N_2 B_2 = B.$$

Since the Duration of this bond portfolio D, therefore,

$$w_1 D_1 + w_2 D_2 = D \Rightarrow \frac{N_1 B_1}{B} D_1 + \frac{N_2 B_2}{B} D_2 = D.$$

Solving for N_1 and N_2, we get,

$$N_1 = \frac{B}{B_1} \left(\frac{D_2 - D}{D_2 - D_1} \right) \text{ and } N_2 = \frac{B}{B_2} \left(\frac{D - D_1}{D_2 - D_1} \right).$$

8.9 Exercise

Exercise 8.1 If the horizon is $h = T - t = 1.7$ years and the annual interest rate is 5%, with the bond price at time t being $B(t) = 98$. Determine the bond price $B(T)$.

Solution: Here, $\mu_{t,T} = 0.05$, $\widetilde{T - t} = 1$ years and $\tau = 0.7$ years. Therefore,

$$B(T) = 98 \times (1 + 0.05)^1 (1 + 0.7 \times 0.05) = 106.5015.$$

Exercise 8.2 If the value of a zero-coupon bond at time $T = 2$ is 100, with the risk-free annual rate being 8%, then determine the price of the bond at time $t = 0$.

Solution: Here, $\mu_f = 0.08$ and $B(2) = 100$. Therefore,

$$B_{0.08}(0) = \frac{100}{(1.08)^2} = 85.7339.$$

Exercise 8.3 If $D_m = 4$ and $d\mu_f = 5\%$ then determine the decrease in the price of a bond, currently selling at 100, and determine the value of the bond as a result of the change in the yield.

Solution: Since $D_m = 4$ and $d\mu_f = 0.05$, therefore $\frac{dB}{B} = -D_m \times d\mu_f = -0.2$ or -20%. This means that as a result of $d\mu_f = 5\%$, the bond price decreases by 20% from 100 to 80.

Exercise 8.4 We create a bond portfolio of two zero-coupon bonds with maturity of 2 years and 4 years, respectively, so that the resulting bond portfolio has a Duration of 3. If w_1 and w_2 are the respective weights of these two bonds, then determine the value of $\frac{w_1}{w_2}$.

Solution: Here, $D_1 = 2$ and $D_2 = 4$. Therefore $D_P = w_1 \times 2 + w_2 \times 4 \Rightarrow 2w_1 + 4(1 - w_1) = 3$. Thus, $w_1 = \dfrac{1}{2} = w_2$. Hence, $\dfrac{w_1}{w_2} = 1$.

Exercise 8.5 Consider the coupon bond with the face value of 100 and annual coupons of 8, with the maturity being 10 years and the annual compounding rate of interest being 6%. Let $\left(\dfrac{dB}{B}\right)_D$ and $\left(\dfrac{dB}{B}\right)_C$ denote the estimates of $\dfrac{dB}{B}$ using only Duration and using both Duration and Convexity, respectively. Then tabulate the values of $\left(\dfrac{dB}{B}\right)_D$ and $\left(\dfrac{dB}{B}\right)_C$ for $d\mu_f = 1, 2, 3, 4$, and 5%.

Solution: From the previous examples, we have $D_m = 7.0236$ and $C = 65.1716$. Then, using the formulas $\left(\dfrac{dB}{B}\right)_D = -D_m d\mu_f$ and $\left(\dfrac{dB}{B}\right)_C = -D_m d\mu_f + \dfrac{1}{2} C(d\mu_f)^2$, we have,

μ_f	$\left(\dfrac{dB}{B}\right)_D$	$\left(\dfrac{dB}{B}\right)_C$
1%	−0.0702	−0.0670
2%	−0.1405	−0.1274
3%	−0.2107	−0.1814
4%	−0.2809	−0.2288
5%	−0.3512	−0.2697

Exercise 8.6 Consider the coupon bond with the face value of 100 and annual coupon of 6, with the maturity being 2 years, and annual compounding rate of interest is 4%. Calculate the price, Duration, and Convexity of the bond.

Solution: Here, $T = 2$, $B(2) = 100$, $c = 6$, and $\mu_f = 0.04$. Therefore,

$$B_{0.04}(0) = \sum_{t=1}^{2} \frac{6}{(1.04)^t} + \frac{100}{(1.04)^2} = 103.7712.$$

$$D = \frac{1}{103.7712} \left[\sum_{t=1}^{2} \frac{6t}{(1.04)^t} + \frac{100 \times 2}{(1.04)^2} \right] = 1.9444.$$

$$C = \frac{1}{103.7712 \times (1.04)^2} \left[\sum_{t=1}^{2} \frac{6t(t+1)}{(1.04)^t} + \frac{100 \times 2 \times 3}{(1.04)^2} \right] = 5.3418.$$

Risk Management of Portfolios

9

In the course of our discussion on portfolio analysis, we have primarily identified variance and semi-variance (or equivalently standard deviation and semi-deviation, respectively) as measures of risk of an asset or a portfolio. While the choice of semi-variance (equivalently semi-deviation) resolves the problem of penalizing "good" returns (observed in case of variance), it however, from an investors' point of view, does not provide any indication of the bounds on the potential losses. Accordingly, this brings into picture the concept of Value-at-Risk (VaR), designed to quantify the losses at given confidence level for a given time horizon. An alternative and more fullproof measure (than VaR) is the Conditional Value-at-Risk (CVaR) which is amenable to indicating potentially extreme losses. Having said so, VaR is the most widely accepted risk measure, used both by financial institutions and regulators, especially in terms of the Basel banking regulations.

9.1 Value-at-Risk

From the perspective of the exposure of risk, a financial institution needs an assessment of the extent to which its portfolio may experience losses in the future. Given the enormously large number of market variables that can impact the risk profile of the portfolio of a financial institution, a simpler and tractable metric of the bank's risk exposure in the immediate future (typically 1 day) for the entire trading portfolio of the bank was needed to be presented to the top management. Accordingly, J.P. Morgan devised, what came to be known as the 4:15 VaR report, through a simple adaptation of the Markowitz framework. The enormous task of setting up a framework of collating vast amounts of information to derive VaR was in place by 1993, followed by VaR rapidly evolving as an important measure of market risk, in the paradigm of the Basel regulations.

S. P. Chakrabarty and A. Kanaujiya, *Mathematical Portfolio Theory and Analysis*,
Compact Textbooks in Mathematics, https://doi.org/10.1007/978-981-19-8544-7_9

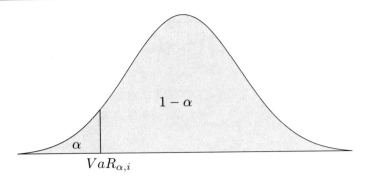

Fig. 9.1 Probability distribution of returns

Definition 9.1.1 The VaR of an asset (or a portfolio) at a confidence level of $100(1 - \alpha)\%$ ($\alpha \in (0, 1)$) for 1-day horizon is that value of the loss after one day, which will not be exceeded in case of $100(1 - \alpha)\%$ of daily return distribution and may exceed in case of at most $100\alpha\%$ of observations of the daily return distribution. Formally, if r_i is the daily return of an asset a_i, then the 1-day VaR (denoted $VaR_{\alpha,i}$, at $100(1 - \alpha)\%$ confidence level) is defined by

$$P(r_i \leq \text{VaR}_{\alpha,i}) = \alpha.$$

Graphically, this is presented in Fig. 9.1 and can now be rewritten as,

$$P(r_i \leq \text{VaR}_{\alpha,i}) = \alpha \iff P\left(\frac{r_i - \mu_i}{\sigma_i} \leq \frac{\text{VaR}_{\alpha,i} - \mu_i}{\sigma_i}\right) = \alpha$$

$$\iff P\left(z \leq \frac{\text{VaR}_{\alpha,i} - \mu_i}{\sigma_i}\right) = \alpha,$$

where, as before, μ_i and σ_i are the expected return and standard deviation, respectively of the (daily) return of the asset a_i. Let z_α be defined as the number that satisfies $P(z \geq z_\alpha) = \alpha$, i.e., $P(z \leq -z_\alpha) = \alpha$. Then,

$$\frac{\text{VaR}_{\alpha,i} - \mu_i}{\sigma_i} = -z_\alpha \Rightarrow \text{VaR}_{\alpha,i} = \mu_i - z_\alpha \sigma_i.$$

In case of the daily returns being normally distributed, $z_{0.05} = 1.645$ and $z_{0.01} = 2.326$, resulting in the 1-day VaR, at 95 and 99% confidence level, being $\text{VaR}_{0.05,i} = \mu_i - 1.645\sigma_i$ and $\text{VaR}_{0.01,i} = \mu_i - 2.326\sigma_i$, respectively. Note that it is customary to consider 1-day VaR at 95 and 99% confidence levels, which motivated the choices of α's, in the preceding narrative. Another customary horizon for calculation of VaR is 10 days, which leads to the question of a T-day VaR. In case of the daily return variables being independent and identically distributed, the T-day VaR is given by,

$$\text{VaR}^T_{\alpha,i} = \mu_i T - z_\alpha \sigma_i \sqrt{T}.$$

Over a small horizon (such as the customary one of 1 day) the value of μ_i is much smaller than σ_i (and also $-z_\alpha \sigma_i$) for the typical values of z_α used, which renders the

1-day and T-day VaR to be $\text{VaR}_{\alpha,i} = -z_\alpha \sigma_i$ and $\text{VaR}^T_{\alpha,i} = -z_\alpha \sigma_i \sqrt{T} = \text{VaR}_{\alpha,i} \sqrt{T}$ (of course assuming that the daily returns are i.i.d). Now, since $\text{VaR}_{\alpha,i}$ (and $\text{VaR}^T_{\alpha,i}$) are values of returns, their negative values would imply the value of loss. Reconciling that VaR is designed to measure loss (negative of return), henceforth, we will take the negative of the VaR as defined above, which results in $\text{VaR}_{\alpha,i} = z_\alpha \sigma_i$ and $\text{VaR}^T_{\alpha,i} = z_\alpha \sigma_i \sqrt{T} = \text{VaR}_{\alpha,i} \sqrt{T}$. Furthermore, the VaR thus defined is in terms of percentage (since it is derived from returns) and upon multiplication by the wealth W_i invested in asset a_i gives the VaR in absolute terms of the currency. Thus, the 1-day VaR in absolute currency is given by,

$$\text{VaR}^c_{\alpha,i} = z_\alpha \sigma_i W_i.$$

9.2 VaR of a Portfolio

Let us consider a portfolio P comprising of n assets whose return is given by $r_p = \sum_{i=1}^n w_i r_i$. Then the portfolio variance is given by $\sigma_P = \mathbf{w} C \mathbf{w}^\top$, and accordingly, the percentage VaR of the portfolio is given by (assuming negligible μ_p compared to σ_P),

$$\text{VaR}_{\alpha,P} = z_\alpha \sigma_P W,$$

where W is the total value of the portfolio. Now,

$$\text{VaR}^2_{\alpha,P} = (z_\alpha \sigma_P W)^2$$

$$= z_\alpha^2 \left(\sum_{i=1}^n w_i^2 \sigma_i^2 + 2 \sum_{i=1}^n \sum_{i<j} w_i w_j \rho_{ij} \sigma_i \sigma_j \right) W^2$$

$$= \sum_{i=1}^n (z_\alpha \sigma_i w_i W)^2 + 2 \sum_{i=1}^n \sum_{i<j} (z_\alpha \sigma_i w_i W)(z_\alpha \sigma_j w_j W) \rho_{ij}$$

$$\leq \sum_{i=1}^n (z_\alpha \sigma_i W_i)^2 + 2 \sum_{i=1}^n \sum_{i<j} (z_\alpha \sigma_i W_i)(z_\alpha \sigma_j W_j)$$

$$= \sum_{i=1}^n \text{VaR}^2_{\alpha,i} + 2 \sum_{i=1}^n \sum_{i<j} \text{VaR}_{\alpha,i} \text{VaR}_{\alpha,j}$$

$$= \left(\sum_{i=1}^n \text{VaR}_{\alpha,i} \right)^2.$$

This gives,

$$\text{VaR}_{\alpha,P} \leq \sum_{i=1}^n \text{VaR}_{\alpha,i}.$$

Note that here $W_i = w_i W$ is the absolute currency amount invested in asset a_i.

Example 9.2.1 Let daily return of asset a_i be independent and identically normally distributed, with the expected return and risk (as given by the standard deviation) being 10% and 4%, respectively. Then determine the 1-day and 10-day VaR, at confidence level 95% (without neglecting the expected return).

Here, $\mu_i = 0.1$, $\sigma_i = 0.04$, and $\alpha = 0.05$. Hence $z_\alpha = z_{0.05} = 1.645$. Therefore,

$$\text{VaR}_{\alpha,i} = \mu_i - z_\alpha \sigma_i = 0.1 - 1.645 \times 0.04 = 0.0342,$$
$$\text{VaR}_{\alpha,i}^T = \mu_i T - z_\alpha \sigma_i \sqrt{T} = 0.1 \times 10 - 1.645 \times 0.04 \times \sqrt{10} = 0.7919.$$

9.3   Decomposition of VaR

A portfolio of a large financial institution is typically very diverse in nature and can be classified in terms of asset classes, business units (within the financial institution), and particular trades. Accordingly, it sometimes is prudent (especially from the point of risk management) to determine the sensitivity of portfolio VaR to various factors, such as marginal VaR, incremental VaR, and component VaR.

1. *Marginal VaR*: Suppose that the absolute currency amount invested in asset a_i of the portfolio P is W_i (instead of a_i, we can also consider an asset class). Then the marginal VaR of asset a_i is defined as,

$$m\text{VaR}_{\alpha,i} = \frac{\partial(\text{VaR}_{\alpha,P})}{\partial W_i}.$$

   In other words, the marginal VaR gives the sensitivity of the portfolio VaR to a unit change (in absolute currency value) of the exposure to asset a_i. If w_i is the weight of asset a_i, in the portfolio P, then,

$$\frac{\partial \sigma_P^2}{\partial w_i} = 2w_i\sigma_i^2 + 2\sum_{\substack{i,j=1 \\ j\neq i}}^n w_j\sigma_{ij} = 2\sum_{j=1}^n w_j\sigma_{ij}$$

$$\Rightarrow 2\sigma_P\frac{\partial \sigma_P}{\partial w_i} = 2\sum_{j=1}^n w_j\text{Cov}(r_i, r_j)$$

$$\Rightarrow \sigma_P\frac{\partial \sigma_P}{\partial w_i} = \text{Cov}\left(r_i, \sum_{j=1}^n w_j r_j\right) = \text{Cov}(r_i, r_P)$$

$$\Rightarrow \sigma_P\frac{\partial \sigma_P}{\partial w_i} = \sigma_{iP} \Rightarrow \frac{\partial \sigma_P}{\partial w_i} = \frac{\sigma_{iP}}{\sigma_P}.$$

   Let β_i is the beta of asset a_i, with respect to portfolio P. Again, neglecting the expected return of the portfolio and the asset a_i, vis-a-vis their standard derivation counterpart, we get,

$$m\,\text{VaR}_{\alpha,i} = \frac{\partial(z_\alpha \sigma_P W)}{\partial(w_i W)}$$

$$= z_\alpha \frac{\partial \sigma_P}{\partial w_i} = z_\alpha \frac{\sigma_i P}{\sigma_P}$$

$$= \frac{\text{VaR}_{\alpha,P}}{\sigma_P W} \frac{\sigma_i P}{\sigma_P}$$

$$= \text{VaR}_{\alpha,P} \left(\frac{\sigma_i P}{\sigma_P^2} \right) \frac{1}{W}$$

$$= \text{VaR}_{\alpha,P} \left(\frac{\beta_i}{W} \right),$$

where $\beta_i = \dfrac{\sigma_i P}{\sigma_P^2}$.

2. *Incremental VaR*: The incremental VaR is formulated so as to capture the effect of the inclusion of an asset (or an asset class) to the original portfolio without it, and it often plays a role when a trade is be considered, so as to asses the additional risk as a result of doing do. Let P be a portfolio, and let the vector of the new position be **np** resulting in the portfolio $\tilde{P} = P + \textbf{np}$. Then incremental VaR is defined as

$$i\,\text{VaR}_{\alpha,\textbf{np}} = \text{VaR}_{\alpha,P+\textbf{np}} - \text{VaR}_{\alpha,P}.$$

Note that the notation $\text{VaR}_{\alpha,P+\textbf{np}}$ is the VaR of the new portfolio $\tilde{P}$. If the number of units of the asset a_i in the vector **np** (the added assets portfolio) is given by np_i for $i = 1, 2, \ldots, m$, then using Taylor series approximation, it can be shown that,

$$i\,\text{VaR}_{\alpha,\textbf{np}} \approx \sum_{i=1}^{m} m\,\text{VaR}_{\alpha,i} \times np_i,$$

provided that np_i is sufficiently small.

3. *Component VaR:* The component VaR of an asset a_i gives the contribution of the asset to the VaR of the portfolio P. The component VaR of asset a_i measures the change in the portfolio VaR, resulting from the removal of the asset a_i from the portfolio, and is given by

$$c\,\text{VaR}_{\alpha,i} = \frac{\partial(\text{VaR}_{\alpha,P})}{\partial W_i} W_i = m\,\text{VaR}_{\alpha,i} \times W_i.$$

Example 9.3.1 Consider a portfolio of value 1000, comprising of two assets, a_i and a_j, with the respective weights being $w_i = 0.4$ and $w_j = 0.6$. Further, the risks (as given by the standard deviation) of the two assets are given by $\sigma_i = 6\%$ and $\sigma_j = 5\%$, with the correlation coefficient of return of the assets being given by $\rho_{ij} = 0.25$. Determine the 1-day VaR of the portfolio, at 95% confidence level.

Here, $W = 1000$, $w_i = 0.4$, $w_j = 0.6$, $\sigma_i = 0.06$, $\sigma_j = 0.05$, $\rho_{ij} = 0.25$, and $\alpha = 0.05 \Rightarrow z_\alpha = 1.645$. Accordingly, the VaR of asset a_i is $z_\alpha \sigma_i W_i = $

$z_\alpha \sigma_i w_i W = 1.645 \times 0.06 \times 0.4 \times 1000 = 39.48$, and the VaR of asset a_j is $z_\alpha \sigma_j W_j = z_\alpha \sigma_j w_j W = 1.645 \times 0.05 \times 0.6 \times 1000 = 49.35$. Also,

$$\sigma_P^2 = w_i^2 \sigma_i^2 + w_j^2 \sigma_j^2 + 2 w_i \sigma_i w_j \sigma_j \rho_{ij}$$
$$= (0.4)^2 (0.06)^2 + (0.6)^2 (0.05)^2 + 2(0.4)(0.06)(0.6)(0.05)(0.25) = 0.001836.$$

This implies, $\sigma_P = \sqrt{0.001836} = 0.0428$, and hence the portfolio VaR is,

$$z_\alpha \sigma_P W = 1.645 \times 0.0428 \times 1000 = 70.406.$$

9.4 Methods for Computing VaR

This section covers some well-established approaches for computing VaR, primarily used in the context of estimating the market risk exposure, which is a key component in the process of capital requirement under Basel regulations.

9.4.1 Historical Simulation Approach

The historical simulation approach is a nonparametric method adopted to estimate VaR. At the heart of this approach lies the technique of using past (historical) data in order to generate scenarios for the future and estimate the VaR. The historical data used pertains to the past data of market variables, impacting the portfolio, the examples of which include interest rate, stock indices, exchange rates, etc.

Suppose that we have collected the data for $T + 1$ days (or time series data points), which we designate as $t = 0, 1, \ldots, T$, with the corresponding value of the market variable being denoted by vm_t, $t = 0, 1, \ldots, T$. If the current time point is T with the corresponding value of the market variable being vm_T, then the random variable for the value of the market variable for the next time point is vm_{T+1}. This random variable takes the T possible values, using the relation,

$$\frac{vm_{T+1}}{vm_T} = \frac{vm_t}{vm_{t-1}} \text{ for } t = 1, 2, \ldots, T.$$

which implies

$$vm_{T+1} = vm_T \times \frac{vm_t}{vm_{t-1}} \text{ for } t = 1, 2, \ldots, T.$$

The T values of vm_{T+1} give its distribution, and the $100(1 - \alpha)\%$ confidence level VaR is determined to be the integer nearest to the αT worst outcome from the distribution of vm_{T+1}. Now, in order to extend this concept to a portfolio affected by K market variables (such as K stock indices) we take into consideration the weights for each, as w_k for $k = 1, 2, \ldots, K$. Further let W be the total amount invested in the portfolio. This means that the absolute amount invested for k-th market variable is $W_k = w_k W$, today $(t = T)$. Let $vm_t^{(k)}$ for $t = 1, 2, \ldots, T$ and $k = 1, 2, \ldots, K$ be the value of the k-th market variable at time t. Then,

$$vm_{T+1}^{(k)} = vm_T^{(k)} \times \frac{vm_t^{(k)}}{vm_{t-1}^{(k)}} = w_k W \times \frac{vm_t^{(k)}}{vm_{t-1}^{(k)}}$$

This gives the market values of the portfolio projected at time $t = T + 1$ to be the random variable.

$$vm_{T+1} = \sum_{k=1}^{K} vm_{T+1}^{(k)} = \sum_{k=1}^{K} w_k W \frac{vm_t^{(k)}}{vm_{t-1}^{(k)}} = \left(\sum_{k=1}^{K} w_k \frac{vm_t^{(k)}}{vm_{t-1}^{(k)}} \right) \times W.$$

This random variable vm_{T+1} takes T possible values, and the $100(1 - \alpha)\%$ confidence level VaR for the portfolio is determined to be integer nearest to the αT worst outcome from the distribution of vm_{T+1}, for the portfolio.

9.4.2 Delta-Gamma Method

The Delta-Gamma Method (also known as the variance–covariance method) is based on the estimates of variance and covariance of the market variable. If v_i is the value of the asset a_i and mv_k is the value of the k-th market variable, then the change in the value of the i-th asset corresponding to change dmv_k in the k-th market variable value is given by,

$$dv_i = \frac{\partial v_i}{\partial mv_k} \times dmv_k = \Delta \, dmv_k.$$

This can be written as,

$$dv_i = (\Delta \times mv_k) \left(\frac{dmv_k}{mv_k} \right),$$

where $\Delta = \frac{\partial v_i}{\partial mv_k}$. Thus $\Delta \times mv_k$ is the change in absolute currency terms of the value of i-th asset, contingent on the percentage change in the market variable. This gives the absolute currency VaR of the asset a_i as,

$$\text{VaR}_{\alpha,i}^{(k)} = z_\alpha (\Delta \times mv_k) \sigma_{mv_k},$$

where σ_{mv_k} is the standard deviation of $rmv_k = \frac{dmv_k}{mv_k}$. This approach of estimating VaR is known as the **Delta-Normal Method**. The corresponding return VaR is given by,

$$\text{VaR}_{\alpha,i}^{(k)} = z_\alpha \Delta \left(\frac{mv_k}{v_i} \right) \sigma_{mv_k}.$$

This approach is suitable in scenarios where the change in the market variable is small. However, in the event of the change being relatively large, the Delta-Normal method exhibits error, which needs to be reduced by the introduction of second-order terms, leading to the **Delta-Gamma Method** as follows:

$$dv_i = \left(\frac{\partial v_i}{\partial mv_k} \right) dmv_k + \frac{1}{2} \left(\frac{\partial^2 v_i}{\partial mv_k^2} \right) (dmv_k)^2 + \cdots \approx \Delta dmv_k + \frac{1}{2} \Gamma (dmv_k)^2,$$

where $\Gamma = \dfrac{\partial^2 v_i}{\partial mv_k^2}$ is the Gamma. This can now be written as,

$$dv_i = (\Delta mv_k) \frac{dmv_k}{mv_k} + \frac{1}{2} \Gamma (mv_k)^2 \frac{dmv_k}{mv_k}^2$$

$$= (\Delta mv_k) r_{mv_k} + \frac{1}{2} \Gamma (mv_k)^2 r_{mv_k}^2,$$

where r_{mv_k} is the return of the market variable. Since the determination of the VaR in this setup for the Delta-Gamma approach is somewhat difficult, this determination in done for the special case of r_{mv_k} being normally distributed with mean zero and variance σ_k^2, the "variance" of dv_i being,

$$\text{Var}(dv_i) = \Delta^2 (mv_k)^2 \sigma_k^2 + \frac{1}{2} \Gamma^2 (mv_k)^4 \sigma_k^4,$$

resulting in the VaR being,

$$\text{VaR}_{\alpha,i}^{(k)} = z_\alpha \left[\Delta^2 \left(\frac{mv_k}{v_k} \right)^2 \sigma_k^2 + \frac{1}{2} \Gamma^2 \left(\frac{mv_k}{\sqrt{v_k}} \right)^4 \sigma_k^4 \right]^{\frac{1}{2}}.$$

We now collate the VaR of the assets to obtain the VaR of the portfolio in both cases as follows: For the sake of brevity we consider only one market factor, i.e., $k = 1$, and drop the index k.

1. *The Delta-Normal Method*: Let v_P denote the value of the portfolio on n assets. Then,

$$dv_P = \sum_{i=1}^{n} \left(\Delta^{(i)} mv^{(i)} \right) \left(\frac{dmv^{(i)}}{mv^{(i)}} \right) = \sum_{i=1}^{n} u_i^{(D)} r_i,$$

where $u_i^{(D)} = \Delta^{(i)} mv^{(i)}$ and $r_i = \dfrac{dmv^{(i)}}{mv^{(i)}}$. Let $\mathbf{u}^{(D)mv} = \left(u_1^{(D)}, u_2^{(D)}, \ldots, u_n^{(D)} \right)^\top$ and $C_{(D)}^{mv}$ be the variance–covariance matrix of r_i's. Then the variance of dv_P is given by

$$\text{Var}(dv_P) = \mathbf{u}^{(D)mv^\top} C_{(D)}^{mv} \mathbf{u}^{(D)mv}.$$

$$\therefore \text{VaR}_{\alpha, P} = z_\alpha \left(\mathbf{u}^{(D)mv^\top} C_{(D)}^{mv} \mathbf{u}^{(D)mv} \right)^{\frac{1}{2}}.$$

2. *The Delta-Gamma Method*: For this approach we define $u_i^{(G)} = \frac{1}{2}\Gamma_i^2\left(mv^{(i)}\right)^2$.
   This gives (for the value of a portfolio P of n assets)

$$dv_P = \sum_{i=1}^{n}\left[u_i^{(D)}r_i + u_i^{(G)}r_i^2\right].$$

Further, let $\mathbf{u}_{(G)}^{mv} = (u_1^{(G)}, u_2^{(G)}, \ldots, u_n^{(G)})^\top$ and $C_{(G)}^{mv}$ be the variance–covariance matrix of r_i^2's. Then the "variance" of the dv_P is given by,

$$\mathrm{Var}(dv_P) = \mathbf{u}^{(D)mv^\top}C_{(D)}^{mv}\mathbf{u}^{(D)mv} + \mathbf{u}^{(G)mv^\top}C_{(G)}^{mv}\mathbf{u}^{(G)mv}.$$

$$\therefore \mathrm{VaR}_{\alpha, P} = z_\alpha\left(\mathbf{u}^{(D)mv^\top}C_{(D)}^{mv}\mathbf{u}^{(D)mv} + \mathbf{u}^{(G)mv^\top}C_{(G)}^{mv}\mathbf{u}^{(G)mv}\right)^{\frac{1}{2}}.$$

9.4.3   Monte Carlo Simulation

A practical approach to estimation of VaR is (akin to many applications in Mathematical Finance) the Monte Carlo simulation. The goal is the estimation of 1-day VaR. This can be presented as an algorithm as follows:

1: **procedure** START
2:    The value v_P of the portfolio today is determined.
3:    Generate sample values of changed dmv_k from a multivariate distribution.
4:    Use the generate sample values in Step 3 for estimaton of each market variable value for the next day.
5:    Use the value from Step 4 to value the estimate for the portfolio value of next day.
6:    Subtract the portfolio value of Step 5 from the portfolio value of Step 2.
7:    Repeat Step 3 to 6 for a large number of times to determine the probability distribution of ΔP.
8:    Calculate VaR contingent on the appropriate percentage of the distribution from Step 7.
9: **end procedure**

Example 9.4.1 Let $-20.05, -16.47, -8.04, -7.09, -6.89, -6.54, -6.11, -5.67,$ $-4.25, -4.01, -3.99, -3.43, -3.23, -2.67,$ and -2.03% be the 15 worst possible distribution of the market values during a period of 1 year (241 days). What will be largest worst possible distribution of the market value at 95% confidence level?

Here, $T = 241$ and $\alpha = 0.05$. Then $\alpha T = 12.05$, and accordingly the 12-th worst distribution of the market value is -3.43%.

Example 9.4.2 Let the values of two assets a_i and a_j be 50 and 60, respectively, with the current value and risk (as given by the variance) of market variable being 10, 000 and 25%, respectively. If the change in the absolute currency terms of the value of asset a_i and a_j, with respect to the market variable are 2% and 3%, respectively, then determine the return VaR of both the assets at 95% confidence level, using Delta-Normal method.

Here $\alpha = 0.05$, $z_\alpha = 1.645$, $v_i = 50$, $v_j = 60$, $mv_k = 10, 000$, $\Delta_i = 0.02$, $\Delta_j = 0.03$, and $\sigma_{mv_k} = 0.05$. Therefore,

$$\text{VaR}_{\alpha,i} = z_\alpha \Delta_i \left(\frac{mv_k}{v_i} \right) \sigma_{mv_k} = \frac{1.645 \times 0.02 \times 10000 \times 0.5}{50} = 3.29,$$

$$\text{VaR}_{\alpha,j} = z_\alpha \Delta_j \left(\frac{mv_k}{v_j} \right) \sigma_{mv_k} = \frac{1.645 \times 0.03 \times 10000 \times 0.5}{60} = 4.1125.$$

9.5   Determination of Volatility

The $\text{VaR}_{\alpha,i}$ (absolutely currency $\text{VaR}_{\alpha,i}$) involves the term $z_\alpha \sigma_i$ ($z_\alpha \sigma_i W_i$). In both cases, the z_α is fixed for all assets and W_i is known (since it is the amount invested in asset a_i). Accordingly, the volatility σ_i plays a vital role (among all the input variables) in the determination of the VaR. Therefore, in this section, we elaborate on some approaches to estimate the volatility.

1. *Daily Volatility*: Let σ_T denote the daily volatility of a market variable estimated at the end of the day $T - 1$. Let mv_t be the value of the market variable at the end of day t. Assuming that the market variable follows the gBm, we define the log -returns as,

$$Rm_t := \ln \left(\frac{mv_t}{mv_{t-1}} \right).$$

If we use the preceding T' values of the market variable, i.e., $mv_{t-1}, t = T, T - 1, \ldots, T - T' + 1$, then the estimate of daily volatility is given by,

$$\widehat{\sigma}_T = \sqrt{\frac{1}{T' - 1} \sum_{t=1}^{T'} (Rm_{T-t} - \widehat{\mu}_T)^2},$$

where $\widehat{\mu}_T = \frac{1}{T} \sum_{t=1}^{T'} Rm_{T-t}$ are both unbiased estimators of σ_T and μ_T, respectively. Recall that we had assumed (at the time of giving an expression for VaR) that the expected return is much smaller than the standard deviation. Further we change the return to

$$rm_t = \frac{mv_t - mv_{t-1}}{mv_{t-1}},$$

and replace the unbiased estimator factor of $T' - 1$, to a maximum likelihood estimator factor of T to obtain,

$$\hat{\sigma}_T = \sqrt{\frac{1}{T'} \sum_{t=1}^{T'} rm_{T-t}^2}.$$

2. *Exponentially Weighted Moving Average (EWMA)*: We begin with a general weight-based approach for estimation of volatility, based on the argument that it would be more realistic to assign greater weight to more recent dataset. Accordingly, we have

$$\sigma_T^2 = \sum_{t=1}^{T'} w_t rm_{T-t}^2.$$

The weight w_t is the weight assigned to the data from observation t days ago. Further $w_t > 0$, $w_t < w_s$, for $t > s$ and $\sum_{t=1}^{T'} w_t = 1$. A particular case of the above is the EWMA, wherein

$$w_{t+1} = \lambda w_t \text{ for } 0 < \lambda < 1.$$

It can be shown that the following recursive relation holds:

$$\sigma_T^2 = \lambda \sigma_{T-1}^2 + (1 - \lambda) rm_{T-1}^2.$$

This leads to,

$$\begin{aligned}
\sigma_T^2 &= \lambda \left[\lambda \sigma_{T-2}^2 + (1 - \lambda) rm_{T-2}^2 \right] + (1 - \lambda) rm_{T-1}^2 \\
&= (1 - \lambda) \left[rm_{T-1}^2 + \lambda rm_{T-1}^2 \right] + \lambda^2 \sigma_{T-2}^2 \\
&= (1 - \lambda) \left[rm_{T-1}^2 + \lambda rm_{T-1}^2 + \lambda^2 rm_{T-3}^2 \right] + \lambda^3 \sigma_{T-3}^2 \\
&= \ldots \\
&= (1 - \lambda) \left(\sum_{t=1}^{T'} \lambda^{t-1} rm_{T-t}^2 \right) + \lambda^{T'} \sigma_{T-T'}^2.
\end{aligned}$$

Since $0 < \lambda < 1$, therefore for large T' the second term $\left(\lambda^{T'} \sigma_{T-T'}^2 \right)$ is negligible compared to the first term, which leads to,

$$\sigma_T^2 = \sum_{t=1}^{T'} (1 - \lambda) \lambda^{t-1} rm_{T-t}^2,$$

which has the weight,

$$w_t = (1 - \lambda) \lambda^{t-1}.$$

The specified properties of w_t can be verified easily.

3. *GARCH*(1, 1) *Model*: The GARCH(1, 1) model is an extension of the EWMA model in the sense of inclusion of the long-term variance rate, var_L, and is given by,

$$\sigma_T^2 = \alpha \, rm_{T-1}^2 + \beta \, \sigma_{T-1}^2 + \gamma \, \text{var}_L,$$

with $\alpha + \beta + \gamma = 1$ (choosing $\alpha = 1 - \lambda, \beta = \lambda$, and $\gamma = 0$ reduces it to EWMA). We now deal with the estimate of volatility as follows:

$$\sigma_T^2 = \alpha \, rm_{T-1}^2 + \beta \, \sigma_{T-1}^2 + (1 - \alpha - \beta) \, \text{var}_L \Rightarrow \sigma_T^2 - \text{var}_L$$
$$= \alpha \left(rm_{T-1}^2 - \text{var}_L \right) + \beta \left(\sigma_{T-1}^2 - \text{var}_L \right).$$

Now we consider the time point $T + T'$, which results in,

$$\sigma_{T+t}^2 - \text{var}_L = \alpha \left(rm_{T+t-1}^2 - \text{var}_L \right) + \beta \left(\sigma_{T+t-1}^2 - \text{var}_L \right).$$

Given the gap of t, the expected value of rm_{T+t-1}^2 is σ_{T+t-1}^2. Hence,

$$E\left[\sigma_{T+t}^2 - \text{var}_L \right] = (\alpha + \beta) E\left[\sigma_{T+t-1}^2 - \text{var}_L \right] = (\alpha + \beta)^t [\sigma_T^2 - \text{var}_L].$$

Therefore,

$$E\left[\sigma_{T+t}^2 \right] = \text{var}_L + (\alpha + \beta)\left[\sigma_T^2 - \text{var}_L \right].$$

This relation gives a mechanism of forecast the volatility, on day $T + t$, based on the information available at the end of day $T - 1$.

Example 9.5.1 The closing prices of a stock are 81.81, 83.07, 83.04, 81.01, 80.75, 81.47, 82.50, 81.95, 79.75, 80.25, and 80.08. Determine the daily volatility, using return and as well as log-returns.

The returns and log-returns are as tabulated below:

Day	Closing Price	Returns (rm_t)	rm_t^2	Log-returns Rm_t	$(Rm_t - \widehat{\mu_T})^2$
1	81.81	0.0154	0.2372	0.0153	0.3036
2	83.07	−0.0004	0.0001	−0.0004	0.0032
3	83.04	−0.0244	0.5976	−0.0247	0.5112
4	81.01	−0.0032	0.0103	−0.0032	0.0012
5	80.75	0.0089	0.0795	0.0089	0.1214
6	81.47	0.0126	0.1598	0.0126	0.2162
7	82.50	−0.0067	0.0444	−0.0067	0.0207
8	81.95	−0.0268	0.7207	−0.0272	0.6286
9	79.75	0.0063	0.0393	0.0063	0.0704
10	80.25	−0.0021	0.0045	−0.0021	0.000
11	80.08	—	—	—	—
		−0.0204	0.0019	−0.0214	0.0027

Using returns, we obtain,

$$\widehat{\sigma_T} = \sqrt{\frac{1}{10}\sum_{t=1}^{10} rm_t^2} = \sqrt{\frac{0.0019}{10}} = \sqrt{0.00019} = 0.0138.$$

Using log-returns, we obtain,

$$\widehat{\mu_T} = \frac{1}{10}\sum_{t=1}^{10} Rm_t = -0.00214,$$

and

$$\widehat{\sigma_T} = \sqrt{\frac{1}{10-1}\sum_{t=1}^{10}(Rm_t - \widehat{\mu_T})^2} = \sqrt{\frac{1}{9} \times 0.0027} = \sqrt{0.0003} = 0.0173.$$

Example 9.5.2 Suppose the returns on a stock for 6 days are $1, 1.4, -0.5, -0.8, 0.7,$ and 1.2%. If $\lambda = 0.9$, then estimate the volatility on the 7-th day.

Here,

$$\sigma_7^2 = \sum_{t=1}^{6}(1-\lambda)\lambda^{t-1} rm_{7-t}^2 = (1-0.9)\left[(0.9)^5(0.01)^2 + (0.9)^4(0.014)^2 + (0.9)^3(-0.005)^2\right.$$
$$+ \ (0.9)^2(-0.008)^2 + (0.9)(0.007)^2 + (0.012)^2\bigg] = 0.000044581.$$

Therefore,

$$\sigma_7 = \sqrt{0.000044581} = 0.0067.$$

9.6 Exercise

Exercise 9.1 Consider a portfolio P (comprising of three assets) with the expected return and risk (as given by the standard deviation) being 1% and 9%, respectively. The amount invested in the portfolio is 1000. Determine the loss to the portfolio, at 99% confidence level, under normal market conditions.

Solution: Here, $W = 1000$, $\mu_P = 0.01$, and $\sigma_P = 0.09$. Also $\alpha = 0.01$ and hence $z_\alpha = z_{0.01} = 2.326$. Therefore,

$$\text{VaR}_{\alpha,P} = z_\alpha \sigma_P W = 2.326 \times 0.09 \times 1000 = 209.34.$$

Exercise 9.2 Consider a portfolio of value 1000, comprising of two asset, a_i and a_j, with the respective weights being $w_i = 0.4$ and $w_j = 0.6$. Further, the risks (as given by the standard deviation) are given by $\sigma_i = 6\%$ and $\sigma_j = 5\%$, and the correlation coefficient of return of the assets is given by $\rho_{ij} = 0.25$. Determine the marginal VaR of each asset at 95% confidence level.

Solution: As seen in the previous Exercise, $\sigma_P^2 = 0.001836$ and therefore $\sigma_P = 0.0428$. Also $\sigma_{iP} = \sigma_P \dfrac{\partial \sigma}{\partial w_i} = w_i \sigma_i^2 + w_j \sigma_i \sigma_j \rho_{ij} = 0.00189$ and $\sigma_{jP} = \sigma_P \dfrac{\partial \sigma}{\partial w_j} = w_j \sigma_j^2$

$+ w_i \sigma_i \sigma_j \rho_{ij} = 0.0018$. Therefore, $\beta_i = \dfrac{\sigma_{iP}}{\sigma_P^2} = \dfrac{0.00189}{0.001836} = 1.0294$ and $\beta_j = \dfrac{\sigma_{jP}}{\sigma_P^2}$

$= \dfrac{0.0018}{0.001836} = 0.9804$. Finally,

$$\text{VaR}_{\alpha,P} = z_\alpha \sigma_P W = 1.645 \times 0.0428 \times 1000 = 70.406,$$
$$m\text{VaR}_{\alpha,i} = \frac{VaR_{\alpha,P} \times \beta_i}{W} = 0.0725,$$
$$m\text{VaR}_{\alpha,j} = \frac{VaR_{\alpha,P} \times \beta_j}{W} = 0.069.$$

Exercise 9.3 Consider the portfolio in the previous Exercise and determine the component VaR of both the assets at 95% confidence level .

Solution: From the previous Exercise, the marginal VaR of asset a_i and a_j are 0.0725 and 0.069, respectively. Therefore

$$c\text{VaR}_{\alpha,i} = m\text{VaR}_{\alpha,i} W_i = m\text{VaR}_{\alpha,i} w_i W = 0.072 \times 0.4 \times 1000 = 29,$$
$$c\text{VaR}_{\alpha,j} = m\text{VaR}_{\alpha,j} W_j = m\text{VaR}_{\alpha,j} w_j W = 0.069 \times 0.6 \times 1000 = 41.4.$$

Exercise 9.4 Consider the portfolio in the previous Exercise and determine the incremental VaR at 95% confidence level by adding an amount 100 of asset a_i and removing 100 of asset a_j.

Solution: From the previous exercise, the marginal VaR of asset a_i and a_j are 0.0725 and 0.069, respectively. Therefore,

$$i\text{VaR}_{\alpha,np} \approx m\text{VaR}_{\alpha,i} \times 100 + m\text{VaR}_{\alpha,j} \times (-100)$$
$$= 0.0725 \times 100 - 0.069 \times 100 = 0.35.$$

Exercise 9.5 Let -20.05, -16.74, -10.04, -9.09, -6.54, -5.67, -4.01, -3.23, -2.67, and -2.03% are the 10 worst possible monthly returns for 10 years (120 months). Determine the largest loss during a month at a 95% confidence level.

Solution: Here, $T = 120$ and $\alpha = 0.05$. Then $\alpha T = 6$. Accordingly, the 6-th worst monthly return is -5.67%.

Exercise 9.6 Let the value of assets a_i and a_j be 50 and 70, respectively, with the market return being normally distributed, with mean 0 and variance 4, respectively. If market variable is 200, then determine the VaR of each asset at 95% confidence level, using the Delta-Gamma method, with $\Delta = 40$ and $\Gamma = 2$.

Solution: Here $\alpha = 0.05 \Rightarrow z_\alpha = 1.645$, $v_i = 50$, $v_j = 70$, $\sigma_k = 2$, $mv_k = 200$, $\Delta = 40$, and $\Gamma = 2$. Therefore,

$$\text{VaR}_{\alpha,1}^{(k)} = z_\alpha \left[\Delta^2 \left(\frac{mv_k}{v_i} \right)^2 \sigma_k^2 + \frac{1}{2} \Gamma^2 \left(\frac{mv_k}{\sqrt{v_i}} \right)^4 \sigma_k^4 \right]^{\frac{1}{2}} = 7463.0,$$

$$\text{VaR}_{\alpha,2}^{(k)} = z_\alpha \left[\Delta^2 \left(\frac{mv_k}{v_j} \right)^2 \sigma_k^2 + \frac{1}{2} \Gamma^2 \left(\frac{mv_k}{\sqrt{v_j}} \right)^4 \sigma_k^4 \right]^{\frac{1}{2}} = 5330.7.$$

Exercise 9.7 Consider a portfolio P of two assets a_1 and a_2, with the tabulated parameters:

Asset (a_i)	σ_i^D	σ_i^G	u_i^D	u_i^G
a_1	0.1000	0.1134	20	40
a_2	0.1044	0.1230	30	50

If the covariance $\sigma_{12}^D = 0.0018$ and $\sigma_{12}^G = 0.0011$, then determine the VaR of portfolio P at confidence level 95%, using the Delta-Gamma method.

Solution: Here, $\alpha = 0.05$ and hence $z_\alpha = 1.645$. Therefore,

$$\mathbf{u}^{(D)} = \begin{bmatrix} 20 \\ 30 \end{bmatrix}, \ \mathbf{u}^{(G)} = \begin{bmatrix} 40 \\ 50 \end{bmatrix}, \ C_{(D)} = \begin{bmatrix} 0.0100 & 0.0018 \\ 0.0018 & 0.0109 \end{bmatrix}, \ C_{(G)} = \begin{bmatrix} 0.0128 & 0.0011 \\ 0.0011 & 0.0151 \end{bmatrix}.$$

Here,

$$\mathbf{u}^{(D)mv^\top} C_{(D)}^{mv} \mathbf{u}^{(D)mv} = 15.97,$$

and

$$\mathbf{u}^{(G)mv^\top} C_{(G)}^{mv} \mathbf{u}^{(G)mv} = 62.63.$$

Therefore,

$$\text{VaR}_{\alpha,P} = z_\alpha \left(\mathbf{u}^{(D)mv^\top} C_{(D)}^{mv} \mathbf{u}^{(D)mv} + \mathbf{u}^{(G)mv^\top} C_{(G)}^{mv} \mathbf{u}^{(G)mv} \right)^{\frac{1}{2}}$$

$$= 1.645 \times \sqrt{15.97 + 62.63} = 1.645 \times 8.8657 = 14.58.$$

Exercise 9.8 Consider a portfolio P of three asset a_1, a_2, and a_3, with the tabulated parameters:
Determine the VaR of portfolio P at 95% confidence level using the Delta-Normal method.

Asset (a_i)	σ_i	Pair (a_i, a_j)	σ_{ij}	u_i^D
a_1	0.1000	(a_1, a_2)	0.0018	20
a_2	0.1044	(a_2, a_3)	0.0011	30
a_3	0.1411	(a_3, a_1)	0.0026	40

Solution: Here, $\alpha = 0.05$ and $z_\alpha = 1.645$. Therefore,

$$\mathbf{u}^{(D)} = \begin{bmatrix} 20 \\ 30 \\ 40 \end{bmatrix} \quad C_{(D)} = \begin{bmatrix} 0.0100 & 0.0018 & 0.0026 \\ 0.0018 & 0.0109 & 0.0011 \\ 0.0026 & 0.0011 & 0.0199 \end{bmatrix}.$$

Hence,

$$\mathrm{VaR}_{\alpha, P} = z_\alpha \left(\mathbf{u}^{(D)mv^\top} C_{(D)}^{mv} \mathbf{u}^{(D)mv} \right)^{\frac{1}{2}} = 1.645 \times \sqrt{54.61} = 12.15.$$

Exercise 9.9 Suppose that the volatility using a GARCH(1, 1) model follows the equation:

$$\sigma_T^2 = 0.00003 + 0.12 rm_{T-1}^2 + 0.70\sigma_{T-1}^2.$$

If the estimated volatility and return on the $(T-1)$ day are 1.2% and 2%, respectively, then determine the estimated volatility on the T day. Further, calculate the volatility per day.

Solution: Here $\alpha = 0.12$, $\beta = 0.70$ then $\gamma = 1 - \alpha - \beta = 1 - 0.12 - 0.70 = 0.18$.

Now $\gamma \times \mathrm{var}_L = 0.00003 \Rightarrow \mathrm{var}_L = \dfrac{0.00003}{0.18} = 0.0001667$.

Therefore, the volatility per day is $\sqrt{0.0001667} = 0.0129$ or 1.29%. Hence, the volatility on the T day is

$$\sigma_T^2 = 0.00003 + 0.12 \times (0.02)^2 + 0.70 \times (0.012)^2 = 0.0001788.$$

Therefore, the estimated volatility on the T day is $\sqrt{0.0001788} = 0.0134$ or 1.34%.

Bibliography

1. N. Amenc, V.L. Sourd, *Portfolio Theory and Performance Analysis* (John Wiley & Sons, 2003)
2. Z. Bodie, A. Kane, A.J. Marcus, P. Mohanty, *Investments* (McGraw-Hill Education, 2018)
3. R.A. Brealey, S.C. Myers, F. Allen, P. Mohanty, *Principles of Corporate Finance* (McGraw-Hill Education, 2018)
4. M. Capinski, T. Zastawniak, *Mathematics for Finance: An Introduction to Financial Engineering* (Springer, 2010)
5. J. Cvitanic, F. Zapatero, *Introduction to the Economics and Mathematics of Financial Markets* (Prentice Hall of India, 2004)
6. E.J. Elton, M.J. Gruber, S.J. Brown, W.N. Goetzmann, *Modern Portfolio Theory and Investment Analysis* (Wiley India, 2009)
7. F.J. Fabozzi, P.N. Kolm, D. Pachamanova, S.M. Focardi, *Robust Portfolio Optimization and Management* (John Wiley & Sons, 2007)
8. E.R. Fernholz, *Stochastic Portfolio Theory* (Springer, 2002)
9. J.C. Francis, D. Kim, *Modern Portfolio Theory: Foundations, Analysis, and New Developments* (Wiley Finance, 2013)
10. O. de La Grandville, *Bond Pricing & Portfolio Analysis: Protecting Investors in the Long Run* (MIT Press, 2003)
11. F.B. Hanson, *Applied Stochastic Processes and Control for Jump Diffusions: Modeling, Analysis and Computation* (Society for Industrial and Applied Mathematics, 2008)
12. J.C. Hull, *Options, Futures and Other Derivatives* (Prentice Hall of India, 2006)
13. J.C. Hull, *Risk Management and Financial Institutions* (Pearson Education, 2011)
14. M.S. Joshi, J.M. Paterson, *Introduction to Mathematical Portfolio Theory* (Cambridge University Press, 2013)
15. D.G. Luenberger, *Investment Science* (Oxford University Press, 1998)
16. B. Oksendal, *Stochastic Differential Equations: An Introduction with Applications* (Springer, 2003)

S. P. Chakrabarty and A. Kanaujiya, *Mathematical Portfolio Theory and Analysis*,
Compact Textbooks in Mathematics, https://doi.org/10.1007/978-981-19-8544-7

17. A.O. Petters, X. Dong, *An Introduction to Mathematical Finance with Applications: Understanding and Building Financial Intuition* (Springer, 2016)
18. J.-L. Prignet, *Portfolio Optimization and Performance Analysis* (Chapman & Hall/CRC Press, 2007)
19. M. Puhle, *Bond Portfolio Optimization* (Springer, 2008)
20. S. Roman, *Introduction to the Mathematics of Finance: From Risk Management to Options Pricing* (Springer, 2004)
21. S. Ross, *Introduction to Probability and Statistics for Engineers and Scientists* (Elsevier, 2009)
22. S. Shreve, *Stochastic Calculus for Finance I: The Binomial Asset Pricing Model* (Springer, 2004)
23. S. Shreve, *Stochastic Calculus for Finance II: Continuous-Time Models* (Springer, 2004)

Printed in the United States
by Baker & Taylor Publisher Services